Oxidation of Primary Alcohols to Carboxylic Acids

Softcover reprint of the hardcover 1st edition 2006

BASIC REACTIONS IN ORGANIC SYNTHESIS

EDITOR-IN-CHIEF: GABRIEL TOJO

DEPARTMENT OF ORGANIC CHEMISTRY

FACULTY OF CHEMISTRY

UNIVERSITY OF SANTIAGO DE COMPOSTELA

15872-SANTIAGO DE COMPOSTELA SPAIN

qogatojo@usc.es

Oxidation of Alcohols to Aldehydes and Ketones:
A Guide to Current Common Practice, by Gabriel Tojo and Marcos Fernández

Oxidation of Primary Alcohols to Carboxylic Acids:
A Guide to Current Common Practice, by Gabriel Tojo and Marcos Fernández

Oxidation of Primary Alcohols to Carboxylic Acids

A Guide to Current Common Practice

GABRIEL TOJO and MARCOS FERNÁNDEZ

 Springer

Authors:
Gabriel Tojo
Department of Organic Chemistry
Faculty of Chemistry
University of Santiago de Compostela
15872-Santiago de Compostela
Spain

Marcos Fernández
Department of Organic Chemistry
Faculty of Chemistry
University of Santiago de Compostela
15872-Santiago de Compostela
Spain

Editor-in-Chief
Gabriel Tojo
Department of Organic Chemistry
Faculty of Chemistry
University of Santiago de Compostela
15872-Santiago de Compostela
Spain

ISBN-13: 978-1-4419-2254-0
e-ISBN-10: 0-387-35432-8
e-ISBN-13: 978-0-387-35432-3

Printed on acid-free paper.

© 2007 Springer Science+Business Media, LLC
Softcover reprint of the hardcover 1st edition 2007

10 9 8 7 6 5 4 3 2 1

springer.com

This book is dedicated to the thousands of scientists cited in the references who constructed our present knowledge on the oxidation of primary alcohols to carboxylic acids. Thanks to their collective effort, the preparation of medicines, pesticides, colorants, and a variety of chemicals that make life more enjoyable, is greatly facilitated.

Preface to the Series

There is natural selection in the synthetic organic laboratory. Successful reagents find their way into specialized journals and tend to populate researchers' benches. Sometimes, old species—like active manganese dioxide in the oxidation of unsaturated alcohols—are so well adapted to a certain reaction niche that they remain unchallenged for a long time. On other occasions, a successful new species—like Dess Martin periodinane—enjoys a population explosion and very quickly inhabits a great number of laboratories. On the other hand, the literature is filled with promising new reagents that fell into oblivion because nobody was able to replicate the initial results on more challenging substrates.

This series, which consists of a collection of monographs on basic reactions in organic synthesis, is not primarily aimed at specialized researchers interested in the development of new reagents. Rather, it is written with the objective of being a practical guide for any kind of scientist, be it a chemist of whatever sort, a pharmacologist, a biochemist, or whoever has the practical need to perform a certain basic synthetic operation in the quickest and most reliable way. Therefore, great emphasis is given to those reagents that are employed most often in laboratories, because their ubiquity proves that they possess a greater reliability. Reagents appearing in only a few publications, regardless of promising potential, are only briefly mentioned. We prefer to err on the side of ignoring some good reagents, rather than including bad reagents that would lead researchers to lose precious time.

The books from this series are meant to be placed near working benches in laboratories, rather than on the shelves of libraries. That is why full experimental details for important reactions are provided. Although many of references from the literature are facilitated, this series is written with the aim of avoiding as much as possible the need to consult original research articles. Many researchers do not have scientific libraries possessing numerous chemical journals readily available, and, many times, although such libraries might be on hand, it is inconvenient to leave the laboratory in order to consult some reference.

Our aim is to facilitate practical help for anybody preparing new organic compounds.

Preface

There is a common view among organic chemists that simple functional group transformations are a mature technology away from the forefront in the Art of Organic Synthesis. This is undoubtedly not the case in the conversion of primary alcohols into carboxylic acids. An ideal reagent for such transformation should be (1) reliable and efficient with regard to all molecules, including complex structures possessing oxidation-sensitive functional groups, (2) cheap, and (3) environmentally friendly. There is no such reagent, even if we limit ourselves to the more mundane need of anything able to provide a certain much-needed carboxylic acid, regardless of price and ecology. This state of affairs is highlighted by the fact that in forty percent of cases the oxidation of primary alcohols to acids is performed using a two-step procedure via the corresponding aldehydes; something that proves that the oxidation of alcohols to aldehydes is a much more mature technology than the oxidation of alcohols to carboxylic acids.

This monograph is a laboratory guide for the transformation of primary alcohols into carboxylic acids. It displays a panorama of the state of the art for this functional group transformation, highlighting the weaponry currently available for scientists and areas where further progress is needed. In conformity with the rest of the series, a selection is made to include those procedures that have proved more reliable in many laboratories around the globe.

Abbreviations

Ac	acetyl	Me	methyl
Alloc	allyloxycarbonyl	MEM	(2-methoxyethoxy)methyl
BAIB	bis(acetoxy)iodobenzene	min.	minute
Bn	benzyl	MOM	methoxymethyl
Boc	*t*-butoxycarbonyl	MS	molecular sieves
b.p.	boiling point	MTBE	methyl *t*-butyl ether
Bu	*n*-butyl	MW	molecular weight
t-Bu	*tert*-butyl	NCS	*N*-chlorosuccinimide
Bz	benzoyl	NMO	*N*-methylmorpholine
ca.	circa		*N*-oxide
CA	*Chemical Abstracts*	NMR	nuclear magnetic
cat.	catalytic		resonance
Cbz	benzyloxycarbonyl	PCC	pyridinium
DCAA	dichloroacetic acid		chlorochromate
DCC	*N*,*N*'-dicyclohexyl	PDC	pyridinium dichromate
	carbodiimide	Ph	phenyl
DMF	dimethylformamide	PhF	9-phenylfluorenyl
DMP	Dess-Martin periodinane	Pht	phthaloyl
DMSO	dimethyl sulfoxide	PMB	*p*-methoxybenzyl
ee	enantiomeric excess	ppm	parts per million
EE	1-ethoxyethyl	Pr	propyl
eq.	equivalent	Py	pyridine
Et	ethyl	ref.	reflux
Fmoc	9-fluorenyl	r.t.	room temperature
	methoxycarbonyl	sat.	saturated
g	gram	T	temperature
h	hour	TBDPS	*t*-butyldiphenylsilyl
i-Pr	isopropyl	TBS	*t*-butyldimethylsilyl
L	liter	TEMPO	2,2,6,6-tetramethyl-
m	multiplet		1-piperidinyloxy free
M	mol/L		radical
MCPBA	*m*-chloroperoxybenzoic	TFA	trifluoroacetic acid
	acid	TFAA	trifluoroacetic anhydride

THF	tetrahydrofuran	TPAP	tetrapropylammonium
THP	tetrahydropyran-2-yl		perruthenate
T_i	internal temperature	v	volume
TIPS	triisopropylsilyl	w	weight
TMS	trimethylsilyl	Z	benzyloxycarbonyl

Contents

Contents

Permanganate

$$O=\overset{\displaystyle O}{\underset{\displaystyle O}{\overset{\|}{\underset{\|}{Mn}}}}\text{-}O^-$$

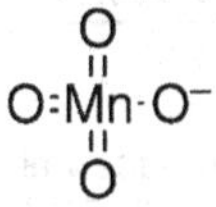

1.1. Introduction

Potassium permanganate is a very strong oxidant that was recurrently employed at the end of the 19th century and the beginning of the 20th century for the determination of organic structures by oxidative degradation. During these degradation studies it became clear that potassium permanganate was able to oxidize primary alcohols to carboxylic acids, and, in 1907[1] and 1909,[2] Fournier proposed the use of aqueous potassium permanganate under strongly alkaline conditions for the oxidation of alcohols to acids. Obviously, these conditions are appropriate for oxidation of alcohols with a certain solubility in water and possessing resistance to degradation by aqueous alkali. In order to broaden the scope of the permanganate oxidation, a number of modifications were later introduced, including:

- Using diverse pH from highly alkaline, as proposed by Fournier, to strongly acidic.[3]
- Adding an organic cosolvent[4] to facilitate the mixing of the alcohol with permanganate.
- Running the reaction in a solvent[5] possessing high solubilizing capacity for both potassium permanganate and alcohols.
- Performing the reaction in a biphasic system with an added phase-transfer catalyst.[6]
- Adding a crown ether[7] that forms a complex with potassium permanganate and facilitates the dissolution of permanganate in organic solvents.
- Employing a tetraalkylammonium permanganate[8] that is soluble in organic solvents thanks to the lipophilic properties of the tetraalkylammonium cation.

The available data are consistent, for most alcohols, with the following mechanism for the oxidation of alcohols to acids using permanganate.

A hydride is transferred—from an alcohol or from the corresponding alkoxide—to either a permanganate anion (MnO_4^-) or to permanganic acid

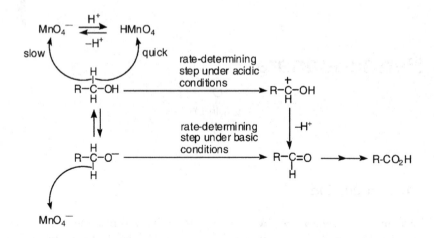

(HMnO$_4$). For obvious electronic reasons, an alkoxide is a much better hydride donor than an alcohol. On the other hand, permanganic acid is a better hydride acceptor than the permanganate anion. In a conjectural situation in which the concentration of starting species played no role, the plausible mechanism would involve a hydride transfer for an alkoxide—the best hydride donor—to permanganic acid—the best hydride acceptor. The pH substantially alters the relative concentration of alcohol versus alkoxide, and permanganic acid versus the permanganate anion, and leads to different pathways depending on proton concentration. Thus, under acidic conditions, in which alkoxide concentration is low and alcohol concentration is high, a hydride is transferred from the alcohol to permanganic acid. Correspondingly, under basic conditions, in which the permanganic acid concentration is low, a hydride is transferred from alkoxide to permanganate anion. This pattern of reactivity causes the oxidation of alcohols with permanganate to be catalyzed both by basic and acidic conditions. Regarding the subsequent oxidation of aldehyde to acid, most experimental evidence shows that this is a very quick transformation that proceeds via aldehyde hydrate, the previous alcohol-to-aldehyde transformation normally being the rate-determining step. This explains the fact that intermediate aldehydes are very rarely isolated from oxidation of primary alcohols with permanganate.

> Nonetheless, some electron-rich aromatic aldehydes possess a certain resistance to oxidation with permanganate under basic conditions,[9] and are occasionally isolated during the oxidation of primary alcohols.[10] Oxidation of benzyl alcohol with permanganate in 20% acetic acid leads to benzaldehyde as the main product.[11]

Information regarding the exact dependence of oxidation speed on pH can be gathered from Figure 1, in which benzhydrol (Ph$_2$CHOH) is oxidized with permanganate at different pH values. This figure represents the oxidation of a secondary alcohol to ketone, rather than a primary alcohol to acid. The

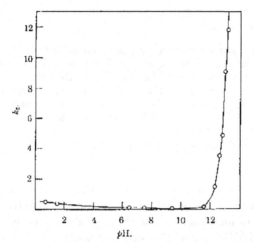

Figure 1. Effect of variation in pH on the bimolecular rate constant, k_2, for the permanganate–benzhydrol reaction; T = 25 °C.[12] Reproduced by permission of the American Chemical Society.

oxidation of both primary and secondary alcohols with permanganate is controlled by a rate-determining hydride transfer. Therefore, it is expected that the kinetic behaviour of the oxidation of primary alcohols parallels the behavior observed during the oxidation of benzhydrol. Figure 1 shows that at a pH ca. 10 there is a minimum of oxidation speed, a small increase of oxidation rate occurring at acidic pH and a drastic increase at very basic pH, particularly above 12.

Figure 1 attests to the convenience of using strongly alkaline conditions for maximum velocity during the oxidation of alcohols with permanganate as originally recommended by Fournier,[1, 2] regardless of the risk of many organic molecules suffering decomposition under strong alkali.

On the other hand, functional groups other than alcohols, such as ethers,[13] can be oxidized by permanganate via mechanisms involving a hydride transfer. As alkoxides are much better hydride donors than alcohols, under strongly basic conditions in which alkoxides exist at a greater concentration, a better selectivity for the oxidation of alcohols relative to other hydride donors is expected. This is another reason why oxidation of alcohols with permanganate must first be tried by default under strongly basic conditions.

> According to ordinary experimental practice, oxidations in which permanganate is brought into an organic solvent by using phase-transfer catalysts or crown ethers, are not normally performed under conditions in which the alcohol is present as an alkoxide, a fact that might lead to a decreased speed and selectivity in the oxidation.

However, useful yields of carboxylic acid can sometimes be obtained employing permanganate under mild basic[14] or acidic[3] conditions, particularly in substrates sensitive to strong alkali.

Research needed

A protocol for oxidation of alcohols in which an alkoxide, formed quantitatively *in situ*, and permanganate ion are dissolved in the same phase must be studied, because it is expected that oxidation of an alkoxide, rather than an alcohol, would guarantee an improved speed and selectivity. A permanganate ion can be brought into an organic solvent by, for instance, an organic counterion, a crown ether, or an organic solvent—such as acetone or pyridine—able to dissolve potassium permanganate.

Normally, the oxidation of alcohols with permanganate is carried out by adding potassium permanganate—either solid or dissolved in water—to a stirred mixture of the alcohol and a highly alkaline aqueous solution prepared by dissolving KOH or NaOH in water. The reaction is normally performed at room temperature, although it is occasionally run at lower[2] temperatures or in boiling[15] water. The success of the oxidation depends critically on the alcohol having a certain solubility in water.[35c] Ideally, the alcohol must be dissolved in the aqueous phase, although the reaction can succeed employing a suspension of the alcohol, provided that some alcohol is dissolved. Sometimes, a cosolvent such as dioxane,[4b, 16] pyridine,[5g] acetone,[17, 5b,f] or *t*-BuOH,[4a, 4c, 18] is added to facilitate the mixing of the alcohol and permanganate in the same phase.

Although the use of strong aqueous alkali is greatly preferred because it promotes speed and selectivity, useful results can occasionally be obtained employing (a) a mild base—such as Na_2CO_3,[14a-c] or NH_3[19]—in water, (b) neutral conditions;[20] or (c) aqueous sulfuric acid.[3]

Although permanganate in strong aqueous alkali provides optimum speed and selectivity in the oxidation of most alcohols, this simple aliphatic alcohol fails to deliver a useful yield of acid using alkaline permanganate, while the oxidation succeeds in aqueous sulfuric acid.[3a]

In order to secure the alcohol and permanganate in the same phase, the oxidation is occasionally performed in an organic solvent, such as acetone,[5a-f] pyridine,[5g,h] or AcOH,[5i-k] in which potassium permanganate possesses a certain solubility. A two phase system consisting of water and an apolar organic solvent can also be used, in which case a phase-transfer catalyst, such as tricaprylylmethylammonium chloride,[6a] *n*-Bu$_4$NBr,[21] *n*-Bu$_4$PCl,[6c] (BnEt$_2$N(+)CH$_2$CH$_2$)$_2$O·2Cl(−),[6b] BnEt$_3$NCl,[6b] Adogen 464,[22] cetrimide,[23] or *n*-Bu$_4$NHSO$_4$,[24] must be added.

Potassium permanganate suffers decomposition in water, resulting in formation of MnO_2 and O_2. This decomposition is strongly accelerated in the presence of acid, base, or MnO_2; the latter being produced not only during the aqueous decomposition of permanganate, but also during the oxidation of alcohols. That is why the decomposition of potassium permanganate can be autocatalytic.[25] Therefore, potassium permanganate must be used in excess during the oxidation of alcohols in the presence of water, especially when strongly basic or acidic conditions are employed.

Zinc permanganate, $Zn(MnO_4)_2$, is a strong oxidizing agent[26] that can explode in contact with organic matter[27] and is occasionally used for the oxidation of primary alcohols to carboxylic acids.[28] In one difficult substrate zinc permanganate was more efficient than ruthenium tetroxide or Jones oxidation, possessing the advantage of maintaining neutral conditions in the reaction medium.[29]

1.2. General Procedure for Oxidation of Primary Alcohols to Carboxylic Acids with Potassium Permanganate

Potassium permanganate (MW 158.04)—either as a ca. 0.2–1 M aqueous solution, or as a finely divided solid—is slowly[a] added to a stirred mixture[b] of 1 equivalent of the alcohol with a ca. 0.1–2.5 N—typically 1 N—NaOH or KOH aqueous solution.[c,d] Approximately 0.4–10 mL of alkali solution per mmol or alcohol are used. The quantity of permanganate employed must be adjusted as the reaction proceeds because it is decomposed with water under acidic, basic, or MnO_2 catalysis at a speed which is difficult to estimate. A sufficient concentration of permanganate must be secured by adding more reagent as long as a substantial quantity of alcohol remains unreacted. Normally, a total of between 1 and 4 equivalents of permanganate must be used. It may be difficult to verify directly the presence of purple permanganate in the solution due to the formation of a thick precipitate of manganese dioxide. The presence of even traces of permanganate can be easily tested by placing a drop of reaction mixture on a piece of filter paper and checking the appearance of a pink ring surrounding the brown spot of manganese dioxide precipitate.[30]

When most of the alcohol is consumed,[e] the excess of potassium permanganate can be optionally destroyed by adding aqueous Na_2SO_3 or an excess of an alcohol such as methanol. The precipitate of manganese dioxide is filtered[f] and the filtrate is optionally washed with an organic solvent such as chloroform, when no risk of extraction of the acid from the alkaline aqueous solution exists. The aqueous solution is acidified by the addition of hydrochloric or sulfuric acid. This sometimes results in the formation of a thick precipitate of the desired organic acid that can be easily separated by filtration. Alternatively, the resulting mixture is extracted with an organic solvent such as ethyl acetate or diethyl ether, and the organic solution is dried (Na_2SO_4 or $MgSO_4$) and concentrated, giving a residue of crude acid that may need further purification.

[a] Normally, the addition is performed over a period between 3 and 30 minutes.

[b] Ideally, the alcohol must be dissolved in the aqueous solution. The reaction can succeed using a suspension of the alcohol provided that some amount is dissolved. The addition of a cosolvent such as dioxane, pyridine, acetone, or t-BuOH can help to bring both the permanganate and the alcohol into the same phase.

[c] The oxidation is catalyzed by both acid and base, being greatly accelerated under a very basic pH, hence the routine use of a 1 N aqueous solution of NaOH or KOH. A very basic pH also guarantees a greater concentration of alkoxide, which is expected to be oxidized with a greater selectivity than an alcohol. Nevertheless, in the case of alcohols with sensitivity to alkali, it may be advisable to perform the reaction under a lower pH. Regardless of decreased velocity and selectivity, useful oxidation yields can sometimes be achieved using a ca. 0.1–1 M— typically 1 M—Na_2CO_3 or K_2CO_3 aqueous solution, plain water or a ca. 1.5–2.5 M H_2SO_4 solution. The oxidation is sometimes performed—with no pH adjustment—in a nonaqueous solvent such as acetone, pyridine, or acetic acid, in which potassium permanganate and most alcohols are soluble.

[d] Normally, the oxidation is carried out at room temperature. The reaction flask is sometimes cooled with a bath at 4–10 °C. For resistant substrates it may be necessary to heat at reflux temperature.

[e] Normally, it takes between 20 minutes and 12 hours.

[f] Sometimes, a substantial quantity of acid is strongly adsorbed on the MnO_2 particles. In such cases, it may be advisable to perform the filtration on a hot mixture or to thoroughly wash the MnO_2 precipitate with plenty of solvent. Boiling a solution containing a suspension of MnO_2 particles may help to coagulate the precipitate and facilitate the filtration. In difficult cases, it may be advisable to dissolve the MnO_2 particles by reduction with sulfur dioxide, either passed through the suspension or generated *in situ* by addition of $NaHSO_3$ and HCl.[25]

$$\text{0.85 eq. } KMnO_4, \text{ NaOH}$$
$$\text{water, } <40\,°C$$

91-93 %

According to the authors: "A number of oxidizing agents were tried in order to ascertain, whether any of them possessed advantages over alkaline permanganate for the oxidation of the benzoylated amino alcohols to the corresponding amino acids. Among those examined and found to react were lead dioxide, sodium dichromate and sulfuric acid, chromic acid, ammonium persulfate, and nitric acid. None of them were as good as alkaline permanganate."[31]

$$\text{1.3 eq. } KMnO_4, Na_2CO_3$$
$$\text{water, 12 h, 0 °C}$$

55%

While $KMnO_4$–KOH, CrO_3–AcOH, and $Na_2Cr_2O_7$–H_2SO_4 produce some degree of racemization during the oxidation to carboxylic acid, potassium permanganate under mild basic conditions secures a reasonable yield of acid without appreciable racemization.[14c]

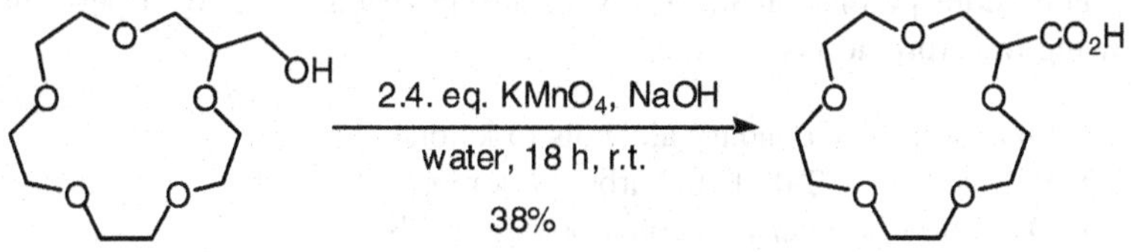

2.4. eq. KMnO$_4$, NaOH

water, 18 h, r.t.

38%

This oxidation failed under a variety of conditions, including Jones oxidation, chromium trioxide in aqueous H$_2$SO$_4$, KMnO$_4$ in benzene and KMnO$_4$ in aqueous H$_2$SO$_4$; while alkaline potassium permanganate provided a useful yield of carboxylic acid.[32]

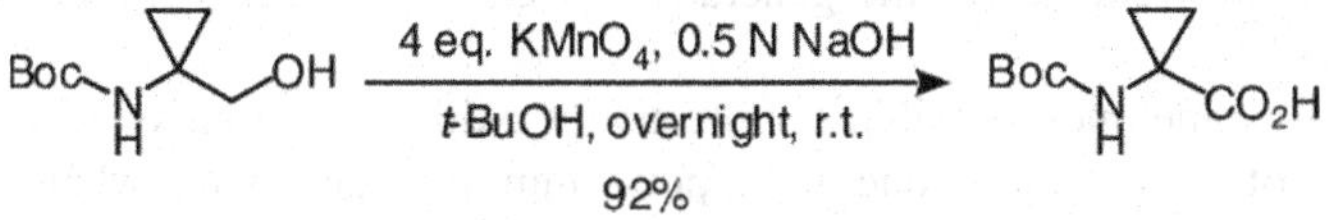

4 eq. KMnO$_4$, 0.5 N NaOH

t-BuOH, overnight, r.t.

92%

While Jones oxidation delivers some aldehyde under mild conditions and side products under more drastic conditions, the use of basic KMnO$_4$ in the presence of t-BuOH as cosolvent allows the isolation of a very good yield of acid.[4a]

1.3. Functional Group and Protecting Group Sensitivity to Potassium Permanganate

Quite expectedly, functional groups with a certain sensitivity to aqueous base, such as esters, are sometimes hydrolyzed during oxidations with basic permanganate.[33] Functional groups with a greater resistance to base, such as urethanes—including the ubiquitous Boc[4c, 34] and Cbz[35] protected amines—remain unchanged during oxidations with basic permanganate. It is sometimes possible to adjust the oxidation conditions to avoid the hydrolysis of esters by using neutral permanganate[36] or acetic acid.[5i, 34a, 37]

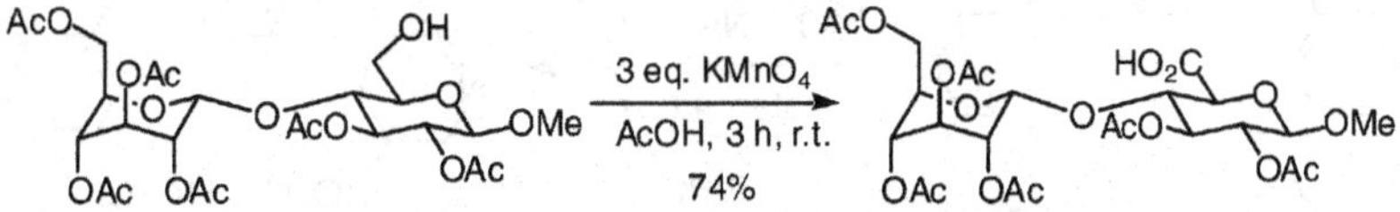

3 eq. KMnO$_4$

AcOH, 3 h, r.t.

74%

This alcohol—containing base-sensitive acetates—can be oxidized to a carboxylic acid with no interference from the acetates, by using potassium permanganate in acetic acid rather than the more common protocol involving permanganate under strong alkaline conditions.[5i]

Potassium permanganate is a very strong oxidant[38] able to perform the following transformations:

- Oxidation of secondary alcohols to ketones.[14e, 39]
- Breakage of 1,2-diols to carbonyl compounds.[35b, 40]
- Oxidation of aldehydes to carboxylic acids.[41]
- Breakage of alkenes to acids,[42] aldehydes,[43] and ketones.[44]
- Oxidation at benzylic positions to aromatic ketones[45] or benzoic acids.[46]
- Conversion of oximes to carbonyl compounds.[47]
- Oxidation of sulfides to sulfones.[48]
- Oxidation of thiols to disulfides.[49]

Because of its very strong oxidizing power, potassium permanganate may not be the best choice for the generation of carboxylic acids in very complex substrates.

It is sometimes possible to selectively oxidize a primary alcohol in the presence of a secondary one with potassium permanganate, when steric or electronic factors play an important role by deactivating the oxidation of the secondary alcohol.[50]

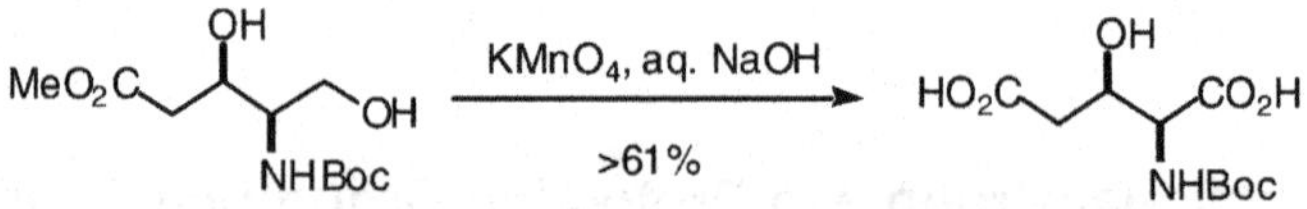

A primary alcohol is selectively oxidized with alkaline potassium permanganate while a secondary alcohol, which is deactivated by neighboring electron-withdrawing groups, remains unreacted. Observe that the basic conditions, quite expectedly, cause the hydrolysis of the methyl ester.[50]

On the other hand, potassium permanganate can be employed to oxidize both primary and secondary alcohols for synthetic convenience.[14e, 51]

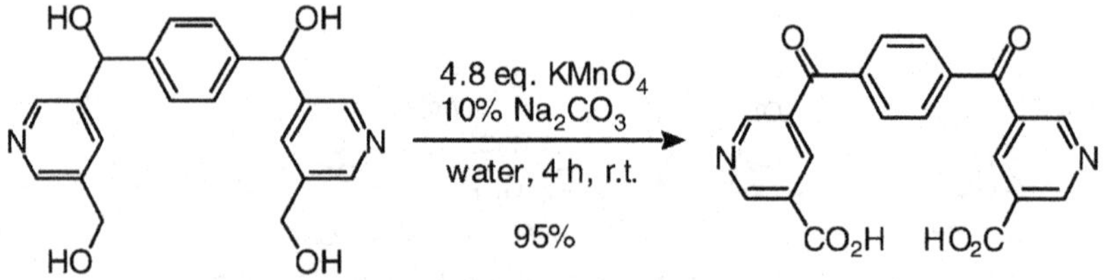

Both primary and secondary alcohols are oxidized by using alkaline potassium permanganate, resulting in an excellent yield of a diketodicarboxylic acid.[14e]

Aldehydes are so easily oxidized by potassium permanganate that it is possible to do so in the presence of unreacting alcohols.[52]

Although alkyl chains linked to aromatic rings can be attacked by potassium permanganate, as these oxidations normally require quite energic conditions, it is usually possible to perform the transformation of a primary alcohol into a carboxylic acid in the presence of aromatic compounds containing alkyl side chains.[6c, 20a, 24, 35c, 53]

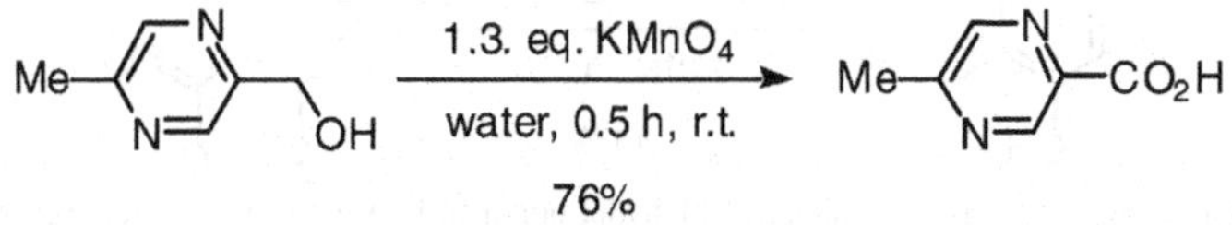

A benzyl alcohol is oxidized with potassium permanganate, in the presence of a methyl group linked to an aromatic ring.[20a]

The selective oxidation of alcohols in the presence of benzylic sites can be done in cases including the ubiquitous benzyl protecting groups,[5a,c] and even the very oxidation-prone p-methoxybenzyl protecting groups.[5f, 18]

Quite interestingly, there are several published instances[19, 54, 55] in which primary amines resist the action of potassium permanganate while a primary alcohol is transformed into a carboxylic acid, which is quite anomalous because permanganate is a very strong oxidant that oxidizes amines very easily.[39b, 49b, 56] It must be mentioned that in all published cases in which alcohols are oxidized in the presence of amines, those are aromatic ones linked to very electron-deficient aromatic rings that render the amine group quite resistant to oxidation.

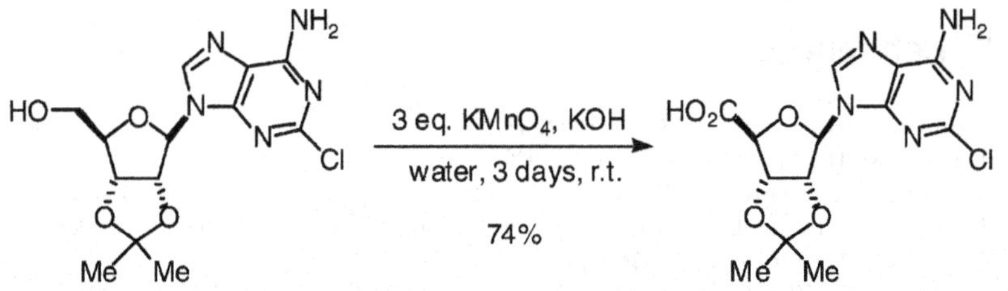

An alcohol is oxidized to a carboxylic acid with basic potassium permanganate in the presence of a primary amine, which is deactivated to oxidation by an electron-deficient purine ring.[54]

1.4. Side Reactions

Sometimes, the intermediate aldehydes are isolated during the oxidation of primary alcohols to carboxylic acids with permanganate.[6c] This side reaction is quite uncommon because aldehydes are normally very quickly transformed into carboxylic acids by permanganate. Exceptionally, some electron-rich aromatic aldehydes are oxidized with permanganate under basic conditions very slowly, resulting in incomplete oxidation of some benzylic alcohols.[10]

This is an unusual case in which an aldehyde, obtained by treatment of an alcohol with potassium permanganate, resists further oxidation. This is probably due to the presence of a phenol at the *ortho* position that strongly deactivates the oxidation of the aldehyde under basic conditions. It is possible to obtain the desired carboxylic acid by employing Ag_2O as oxidant.[10] Quite probably, a permanganate oxidation under neutral or acidic conditions would allow the direct obtention of carboxylic acid.[9]

Racemization at the α-position sometimes occurs during the formation of carboxylic acids. Apparently, this is caused by the strong alkaline conditions normally used in oxidations with potassium permanganate, because racemization can be avoided by employing mildly basic[14c] or acidic[3b] conditions. In fact, potassium permanganate can be the oxidant of choice[14c] when racemization must be avoided (see second example on page 6).

1.5. References

1 Fournier, M. H.; *Compt.Rend.Chim.* **1907,** 331.

2 Fournier, M. H.; *Bull.Soc.Chim.Fr.* **1909,** 920.

3 a) Crombie, L.; Harper, S. H.; *J.Chem.Soc.* **1950,** 2685. b) Fukumasa, M.; Furuhashi, K.; Umezawa, J.; Takahashi, O.; Hirai, T.; *Tetrahedron Lett.* **1991,** *32,* 1059. c) Collins, J. F.; McKervey, M. A.; *J.Org.Chem.* **1969,** *34,* 4172.

4 a) Kordes, M.; Winsel, H.; de Meijere, A.; *Eur.J.Org.Chem.* **2000,** 3235. b) Ambroise, Y.; Pillon, F.; Mioskowski, C.; Valleix, A.; Rousseau, B.; *Eur.J.Org.Chem.* **2001,** 3961. c) Bertus, P.; Szymoniak, J.; *J.Org.Chem.* **2002,** *67,* 3965.

5 a) Kitazume, T.; Yamazaki, T.; Ito, K.; *Synth.Commun.* **1990,** *20,* 1469. b) Epsztajn, J.; Jóźwiak, A.; Szcześniak, A. K.; *Synth.Commun.* **1994,** *24,* 1789. c) Posner, G. H.; Bae Jeon, H.; Ploypradith, P.; Paik, I.-H.; Borstnik, K.; Xie, S.; Shapiro, T. A.; *J.Med.Chem.* **2002,** *45,* 3824. d) Jones, R. G.; *J.Am.Chem.Soc.* **1949,** *71,* 383. e) Gilman, H.; Brannen, C. G.; Ingham, R. K.; *J.Am.Chem.Soc.* **1956,** *78,* 1689. f) Vlahova, R.; Krikorian, D.; Spassov, G.; Chinova, M.; Vlahov, I.; Parushev, S.; Snatzke, G.; Ernst, L.; Kieslich, K.; Abraham, W.-R.; Sheldrick, W. S.; *Tetrahedron* **1989,** *45,* 3329. g) El Hajj, T.; Masroua, A.; Martin, J.-C.; Descotes, G.; *Bull.Soc.Chim.Fr.* **1987,** 855. h) Benkeser, R. A.; Schroeder, W.; *J.Am.Chem.Soc.* **1958,** *80,* 3314.

i) Bock, K.; Pedersen, H.; *Acta Chem.Scand. B* **1988,** *42,* 75. j) Nguyen, T.; Wakselman, C.; *Synth.Commun.* **1990,** *20,* 97. k) Chiu, C. K.-F.; *Tetrahedron: Asymmetry* **1995,** *6,* 881.

6 a) Herriott, A. W.; Picker, D.; *Tetrahedron Lett.* **1974,** *15,* 1511. b) Álvarez-Builla, J.; Vaquero, J. J.; García Navio, J. L.; Cabello, J. F.; Sunkel, C.; Fau de Casa-Juana, M.; Dorrego, F.; Santos, L.; *Tetrahedron* **1990,** *46,* 967. c) Panetta, C. A.; Garlick, S. M.; Durst, D.; Longo, F. R.; Ward, J. R.; *J.Org.Chem.* **1990,** *55,* 5202.

7 a) Sam, D. J.; Simmons, H. E.; *J.Am.Chem.Soc.* **1972,** *94,* 4024. b) Nomura, E.; Taniguchi, H.; Otsuji, Y.; *Bull.Chem.Soc.Jpn.* **1994,** *67,* 309.

8 a) Sala, T.; Sargent, M. V.; *J.Chem.Soc. Chem.Commun.* **1978,** 253. b) Ogawa, Y.; Hosaka, K.; Chin, M.; Mitsuhashi, H.; *Synth.Commun.* **1992,** *22,* 315.

9 Wiberg, K.; Stewart, R.; *J.Am.Chem.Soc.* **1955,** *77,* 1786.

10 See, for example: Sharma, G. V. M.; Ilangovan, A.; Lavanya, B.; *Synth.Commun.* **2000,** *30,* 397.

11 Banerji, K. K.; Nath, P.; *Bull.Chem.Soc.Jpn.* **1969,** *42,* 2038.

12 Stewart, R.; *J.Am.Chem.Soc.* **1957,** *79,* 3057.

13 Barter, R. M.; Littler, J. S.; *J.Chem.Soc. (B)* **1967,** 205.

14 a) Zimmerman, H. E.; English Jr., J.; *J.Am.Chem.Soc.* **1954,** *76,* 2285. b) Blicke, F. F.; Cox, R. H.; *J.Am.Chem.Soc.* **1955,** *77,* 5403. c) Begley, M. J.; Crombie, L.; Jones, R. C. F.; Palmer, C. J.; *J.Chem.Soc. Perkin Trans. I* **1987,** 353. d) Katagiri, N.; Haneda, T.; Watanabe, N.; Hayasaka, E.; Kaneko, C.; *Chem.Pharm.Bull.* **1988,** *36,* 3867. e) Gran, U.; *Tetrahedron* **2003,** *59,* 4303.

15 See, for example: Itazaki, H.; Kawasaki, A.; Matsuura, M.; Ueda, M.; Yonetani, Y.; Nakamura, M.; *Chem.Pharm.Bull.* **1988,** *36,* 3387.

16 Nassal, M.; *Lieb.Ann.Chem.* **1983,** 1510.

17 Oi, S.; Matsunaga, K.; Hattori, T.; Miyano, S.; *Synthesis* **1993,** 895.

18 Rodríguez, M. J.; Belvo, M.; Morris, R.; Zeckner, D. J.; Current, W. L.; Sachs, R. K.; Zweifel, M. J.; *Bioorg.Med.Chem.Lett.* **2001,** *11,* 161.

19 Harper, P. J.; Hampton, A.; *J.Org.Chem.* **1970,** *35,* 1688.

20 a) Pitrè D.; Boveri, S.; Grabitz, E. B.; *Chem.Ber.* **1966,** *99,* 364. b) Miyano, K.; Koshigoe, T.; Sakasai, T.; Hamano, H.; *Chem.Pharm.Bull.* **1978,** *26,* 1465.

21 Arts, N. B. M.; Klunder, A. J. H.; Zwanenburg, B.; *Tetrahedron* **1978,** *34,* 1271.

22 Fahrni, C. J.; Pfaltz, A.; *Helv.Chim.Acta* **1998,** *81,* 491.

23 Coxon, G. D.; Knobl, S.; Roberts, E.; Baird, M. S.; Al Dulayymi, J. R.; Besra, G. S.; Brennan, P. J.; Minnikin, D. E.; *Tetrahedron Lett.* **1999,** *40,* 6689.

24 Sartori, A.; Casnati, A.; Mandolini, L.; Sansone, F.; Reinhoudt, D. N.; Ungaro, R.; *Tetrahedron* **2003,** *59,* 5539.

25 Fieser and Fieser; *Reagents for Organic Synthesis,* Vol. 1; Wiley: New York, **1967,** p. 942–43.

26 Wolfe, S.; Ingold, C. F.; *J.Am.Chem.Soc.* **1983,** *105,* 7755.

27 Soergel, U. C.; *Pharm.Prax.* **1960,** *2,* 30.

28 Misiti, D.; Zappia, G.; Delle Monache, G.; *Gazz.Chim.Ital.* **1995,** *125,* 219.

29 Rajashekhar, B.; Kaiser, E. T.; *J.Org.Chem.* **1985,** *50,* 5480.

30 Fieser and Fieser; *Reagents for Organic Synthesis,* Vol. 1; Wiley, New York, **1967,** p. 943.

31 Billman, J. H.; Parker, E. E.; *J.Am.Chem.Soc.* **1944,** *66,* 538.

32 Takahashi, T.; Mallen, J.; Lorenzo, D.; Khanna, R. K.; Nishikawa, M.; Gokel, G. W.; *Bull.Chem.Soc.Jpn.* **1990,** *63,* 383.

33 Kerber, K.-H.; Lackner, H.; *Lieb.Ann.Chem.* **1989,** 719.

34 a) Holladay, M. W.; Salituro, F. G.; Rich, D. H.; *J.Med.Chem.* **1987,** *30,* 374. b) Ciufolini, M. A.; Swaminathan, S.; *Tetrahedron Lett.* **1989,** *30,* 3027.

35 a) Szirtes, T.; Kisfaludy, L.; Palosi, E.; Szporny, L.; *J.Med.Chem.* **1986,** *29,* 1654. b) Hanessian, S.; Vanasse, B.; *Can.J.Chem.* **1987,** *65,* 195. c) Kashima, C.; Harada, K.; Fujioka, Y.; Maruyama, T.; Omote, Y.; *J.Chem.Soc. Perkin Trans. I* **1988,** 535. d) King, F. D.; Martin, R. T.; *Tetrahedron Lett.* **1991,** *32,* 2281.

36 Subramanian, G. B. V.; Malathi, N.; *Ind.J.Chem.* **1989,** *28B,* 806.

37 Hirasaka, Y.; Matsunaga, I.; *Chem.Pharm.Bull.* **1965,** *13,* 176.

38 Fatiadi, A. J.; *Synthesis* **1987,** 85.

39 a) Hajipour, A. R.; Mallakpour, S. E.; Imanzadeh, G.; *Chem.Lett.* **1999,** *2,* 99. b) Noureldin, N. A.;
    Bellegarde, J. W.; *Synthesis* **1999,** 939. c) Kuo, C.-W.; Fang, J.-M.; *Synth.Commun.* **2001,** *31,* 877.

40 a) Medina, E.; Vidal-Ferrán, A.; Moyano, A.; Pericas, M. A.; Riera, A.; *Tetrahedron: Asymmetry*
    1997, *8,* 1581. b) Macías, F. A.; Aguilar, J. M.; Molinillo, J. M. G.; Rodríguez-Luis, F.;
    Collado, I. G.; Massanet, G. M.; Fronczek, F. R.; *Tetrahedron* **2000,** *56,* 3409. c) Medina, E.;
    Moyano, A.; Pericas, M. A.; Riera, A.; *Helv.Chim.Acta* **2000,** *83,* 972.

41 a) Tashiro, M.; Yamato, T.; *J.Org.Chem.* **1983,** *48,* 1461. b) Wasserman, H. H.; Rodriques, K.;
    Kucharczyk, R.; *Tetrahedron Lett.* **1989,** *30,* 6077. c) Palmer, B. D.; Boyd, M.; Denny, W. A.;
    J.Org.Chem. **1990,** *55,* 438. d) Wright, R. S.; Vinod, T. K.; *Tetrahedron Lett.* **2003,** *44,* 7129.

42 a) Hurd, C. D.; McNamee, R. W.; Green, F. O.; *J.Am.Chem.Soc.* **1939,** *61,* 2979. b) Alder, K.;
    Schneider, S.; *Lieb.Ann.Chem.* **1936,** *524,* 189. c) Birch, S. F.; Oldham, W. J.; Johnson, E. A.;
    J.Chem.Soc. **1947,** 818.

43 Wiberg, K. B.; Saegebarth, K. A.; *J.Am.Chem.Soc.* **1957,** *79,* 2822.

44 a) Crieggee, R.; Rustaiyan, A.; *Chem.Ber.* **1975,** *108,* 749. b) Friedrichsen, W.; Schröer, W.-D.;
    Schmidt, R.; *Lieb.Ann.Chem.* **1976,** 793.

45 a) Lai, S.; Lee, D. G.; *Tetrahedron* **2002,** *58,* 9879. b) Shaabani, A.; Teimouri, F.; Lee, D. G.;
    Synth.Commun. **2003,** *33,* 1057.

46 a) Choi, K.; Hamilton, A. D.; *J.Am.Chem.Soc.* **2003,** *125,* 10241. b) Zhao, X.; Wang, X.-Z.;
    Jiang, X.-K.; Chen, Y.-Q.; Li, Z.-T.; Chen, G.-J.; *J.Am.Chem.Soc.* **2003,** *125,* 15128.

47 a) Jadhav, V. K.; Wadgaonkar, P. P.; Joshi, P. L.; Salunkhe, M. M.; *Synth.Commun.* **1999,** *29,*
    1989. b) Chrisman, W.; Blankinship, M. J.; Taylor, B.; Harris, C. E.; *Tetrahedron Lett.* **2001,** *42,*
    4775. c) Imanzadeh, G. H.; Hajipour, A. R.; Mallakpour, S. E.; *Synth.Commun.* **2003,** *33,* 735.

48 a) Ram, V. J.; Pandey, H. N.; *Bull.Soc.Chim.Belges* **1977,** *86,* 399. b) Basile, M.; Previtera, T.;
    Vigorita, M. G.; Fenech, G.; Pizzimenti, F. C.; *Farmaco* **1986,** *41,* 637. c) Previtera, T.; Basile, M.;
    Vigorita, M. G.; Fenech, G.; Occhiuto, F.; Circosta, C.; Costa de Pasquale, R.; *Eur.J.Med.Chem.*
    1987, *22,* 67.

49 a) Kopecky, J.; Smejkal, J.; *Collect.Czech. Chem.Commun.* **1980,** *45,* 2976. b) Shaabani, A.;
    Lee, D. G.; *Tetrahedron Lett.* **2001,** *42,* 5833. c) Boyd, D. R.; Sharma, N. D.; King, A. W. T.;
    Shepherd, S. D.; Allen, C. C. R.; Holt, R. A.; Luckarift, H. R.; Dalton, H.; *Org.Biomol.Chem.*
    2004, *2,* 554.

50 Garner, P.; *Tetrahedron Lett.* **1984,** *25,* 5855.

51 Lal, K.; Salomon, R. G.; *J.Org.Chem.* **1989,** *54,* 2628.

52 a) Smith, C. W.; Norton, D. G.; Ballard, S. A.; *J.Am.Chem.Soc.* **1951,** *73,* 5273. b) Barber, G. W.;
    Ehrenstein, M.; *J.Org.Chem.* **1961,** *26,* 1230. c) Pichette, A.; Liu, H.; Roy, C.; Tanguay, S.;
    Simard, F.; Lavoie, S.; *Synth.Commun.* **2004,** *34,* 3925.

53 Truesdale, E. A.; Cram, D. J.; *J.Org.Chem.* **1980,** *45,* 3974.

54 Olsson, R. A.; Kusachi, S.; Thompson, R. D.; Ukena, D.; Padgett, W.; Daly, J. W.; *J.Med.Chem.*
    1986, *29,* 1683.

55 Hutchison, A. J.; Williams, M.; De Jesus, R.; Yokoyama, R.; Oei, H. H.; Ghai, G. R.; Webb,
    R. L.; Zoganas, H. C.; Stone, G. A.; Jarvis, M. F.; *J.Med.Chem.* **1990,** *33,* 1919.

56 a) Venkov, A. P.; Statkova-Abeghe, S. M.; *Tetrahedron* **1996,** *52,* 1451. b) Seebacher, W.;
    Brun, R.; Saf, R.; Kroepfl, D.; Weis, R.; *Monat.Chem.* **2004,** *135,* 313. c) Tietze, L. F.;
    Rackelmann, N.; Mueller, I.; *Chem.Eur.J.* **2004,** *10,* 2722.

2

Jones and Other CrO₃-Based Oxidations

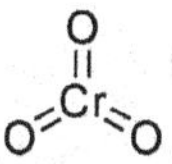

2.1. Introduction

Chromium trioxide (**1**) is a red solid with a very strong oxidizing power that can explode in contact with organic matter.[1]

WARNING: because of the danger of explosion, assays must conform to well-tested protocols during oxidations with CrO₃. Experimental modifications involving less common solvents or temperatures higher than recommended can be particularly dangerous.

Chromium trioxide reacts with water under acidic conditions resulting in an equilibrating mixture containing chromic acid (**2**) and oligomers thereof.[2]

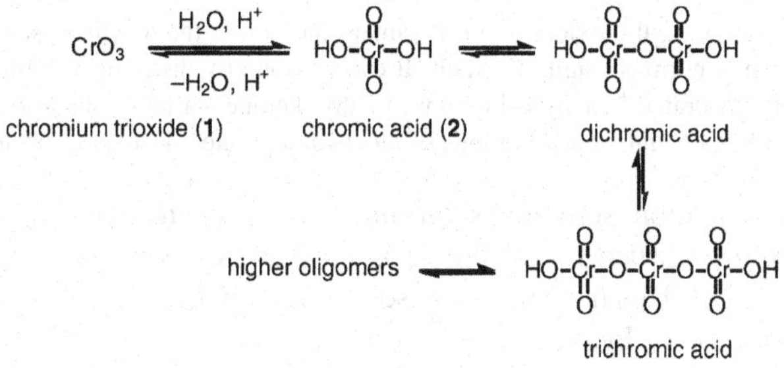

In 1946, Jones et al. reported[3] that secondary alcohols can be oxidized to ketones by pouring an aqueous solution of CrO₃ containing sulfuric acid—often described as either Jones reagent or chromic acid solution—over the alcohol in acetone. Shortly after,[4] Jones et al. showed that this oxidation protocol is also useful for the transformation of primary alcohols into carboxylic acids.

Mechanism

The available experimental data[5] are consistent with a mechanism for oxidation of primary alcohols to acids with Jones reagent, involving the initial formation of

13

a chromate ester (**3**) that evolves to an intermediate aldehyde (**4**). This aldehyde equilibrates with its hydrate (**5**) that reacts resulting in the formation of a new chromate ester (**6**), which decomposes to a carboxylic acid (**7**).

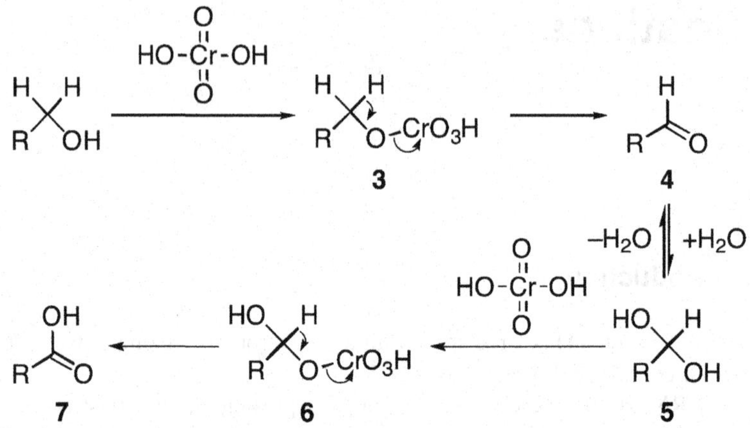

The oxidation of the aldehyde to the corresponding carboxylic acid via hydrate **5** is normally quicker than the oxidation of the starting alcohol to the aldehyde. Therefore, erosion of yield of carboxylic acid due to oxidation being partially suspended at the aldehyde level is usually not a matter of concern.

> In fact, stopping the oxidation of a primary alcohol at the aldehyde stage using Jones oxidation is normally quite difficult. It can be done by distilling volatile aldehydes as they are generated,[6] or by using ethyl methyl ketone—a less polar solvent that averts the building of high concentrations of aldehyde hydrate—instead of acetone.[7]

Jones oxidation sometimes terminates at the aldehyde level using the standard protocol, when the aldehyde is hindered or belongs to a benzylic or allylic kind.[5c, 8, 10] This happens because these aldehydes tend to equilibrate with a small proportion of hydrate.[9]

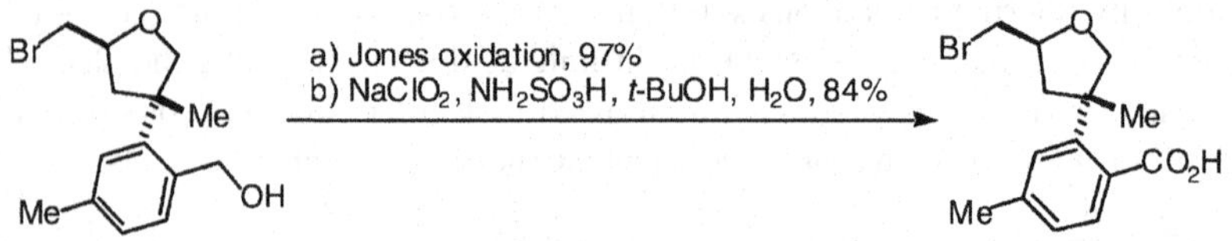

a) Jones oxidation, 97%
b) NaClO$_2$, NH$_2$SO$_3$H, t-BuOH, H$_2$O, 84%

This is a rare case in which a Jones oxidation finishes at the aldehyde stage. Consequently, it is necessary to perform an additional oxidation with NaClO$_2$ to obtain the desired carboxylic acid.[10] The resistance of the aldehyde to react under Jones conditions is most probably due to its benzylic and hindered nature, which is responsible for the existence of a very small percentage of aldehyde hydrate.

Inverse Addition

A common side reaction during the chromic acid oxidation of primary alcohols is the formation of dimeric esters.[11] Craig and Horning proved[12] that this side reaction, already observed in 1953,[13] proceeds via an intermediate hemiacetal, as shown in the equation below.

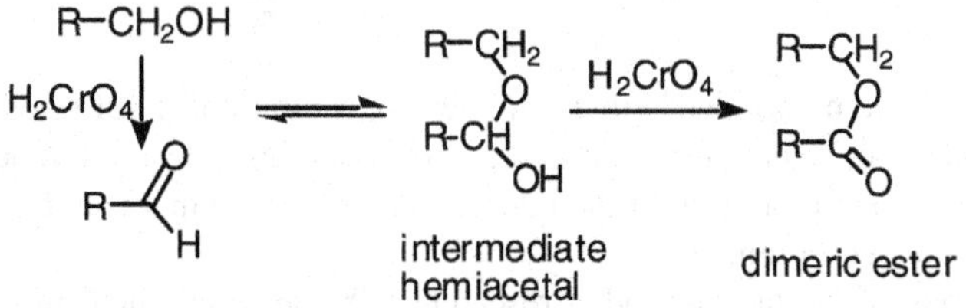

R–CH$_2$OH → R–CHO (H$_2$CrO$_4$) ⇌ intermediate hemiacetal → dimeric ester

Under experimental conditions promoting a high concentration of hemi-acetal, it is possible to obtain very good yields of dimeric esters using Jones reagent.

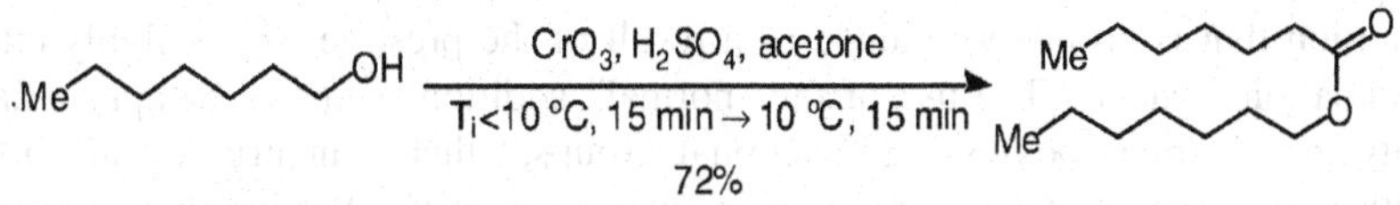

$$\text{Me}\!\!\!\sim\!\!\!\text{OH} \xrightarrow[\text{T}_i<10\,°C,\ 15\ \text{min}\ \to\ 10\,°C,\ 15\ \text{min}]{\text{CrO}_3,\ \text{H}_2\text{SO}_4,\ \text{acetone}}$$

72%

A 72% yield of dimeric ester is obtained by dropping chromic acid over a concentrated solution of heptanol. Under these conditions a high proportion of intermediate hemiacetal builds up during the oxidation.[12]

In 1974, Holland and Gilman proved[14] that using an "inverse addition" procedure allows much better yield of carboxylic acids due to a much lower concentration of hemiacetal formed in the medium. Thus, rather than pouring a chromic acid solution over a solution of the alcohol, it is the solution of the alcohol in acetone that is slowly added over the chromic acid solution. This causes a quick transformation of alcohol—in contact with great excess of chromic acid—into carboxylic acid. This prevents buildup of a high concentration of intermediate aldehyde in contact with starting alcohol and therefore avoids the formation of substantial amounts of hemiacetal. As expected, experimental factors favoring quick oxidation of the alcohol and low concentration of inter-mediate aldehyde lead to greater yield of carboxylic acid.

Thus, as shown in the scheme below, an increased yield of carboxylic acid can be achieved by the following experimental conditions, which promote low concentration of intermediate aldehyde:

- Low concentration of starting alcohol in acetone.
- High concentration of sulfuric acid in the chromic acid solution.
- Long addition time of the alcohol solution.

Me—(CH₂)₅—≡≡—CH₂-CH₂OH $\xrightarrow[\text{acetone}]{\text{1M CrO}_3,\text{ H}_2\text{SO}_4}$ Me—(CH₂)₅—≡≡—CH₂-CO₂H +

Concentration Alcohol/Acetone	Acid Strength	Addition Time	% Yield Acid	Ester
1.0 M	3 N	4 h	–	>50
1.0 M	10 N	2 h	18	50
0.1 M	10 N	7 h	73	12

It must be mentioned[15] that a higher concentration of sulfuric acid in the chromic acid solution causes an increase in oxidizing power of Jones reagent, leading to quicker oxidation of the intermediate aldehyde that is not able to build up a high concentration.

In spite of the fact that other researchers[16] confirmed the efficiency of the "inverse addition" protocol advocated by Holland and Gilman, nowadays, the obtention of carboxylic acids by Jones oxidation is customarily done using the "normal" addition procedure as originally described by Jones *et al.* We recommend employing the "inverse addition" protocol as the default procedure for Jones oxidation to carboxylic acids, rather than using an experimental tradition that is probably retained as a result of the prestige of the highly cited foundational papers. The use of the "normal" addition seems to be appropriate only in substrates possessing functional groups, other than primary alcohols, with a high sensitivity to oxidation. In such cases, it is not advisable to let the substrate be in contact with great excess of oxidizing agent.

Solvent

In the vast majority of cases Jones oxidation is carried out by mixing a chromic acid solution with a solution of the alcohol in *acetone* between 0°C and room temperature.

Very rarely, solvents other than acetone, such as ethyl methyl ketone,[13] acetone:benzene (1:1),[17] acetone:AcOH (2:1),[18] THF,[19] or Et₂O,[20] are used.
Primary alcohols are sometimes transformed into carboxylic acids using CrO₃ or K₂Cr₂O₇[22] in AcOH[23] or pyridine.[24] When AcOH is used, sometimes, it is mixed with water[22a,b, 23c,h, 25] or sulfuric acid.[22, 25a]

Occasionally, less common experimental protocols offer the best results.[19]

3M Jones reagent

Celite® , THF, 1 h, 0 °C

78%

This Jones oxidation is carried out in a solution of starting compound in THF containing Celite®. These uncommon experimental conditions avoid problems associated with α-epimerization and ketal hydrolysis.[19]

Acidity

An obvious limitation of Jones oxidation is the presence of sulfuric acid that may cause reactions on acid-sensitive functionalities and protecting groups. Due to the biphasic nature of Jones oxidation, the contact between the organic phase containing the substrate and the aqueous phase containing sulfuric acid can be quite limited. This allows the survival of protecting groups, such as those present in THP ethers and Boc-protected amines that normally would be cleaved in the presence of acid. Moreover, the quantity of sulfuric acid can be decreased—at the cost of causing certain reduction in the oxidizing power of chromic acid—in order to preserve acid-sensitive moieties.[26]

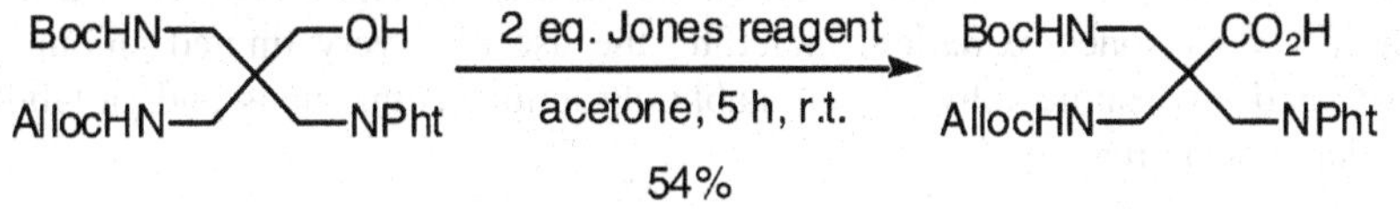

In this oxidation, the concentration of sulfuric acid was carefully tuned to avoid any cleavage of the Boc protection while maintaining the oxidizing power of chromic acid.[26]

On the other hand, the acidity of Jones reagent can be employed for synthetic advantage to perform additional transformations *in situ*.[27]

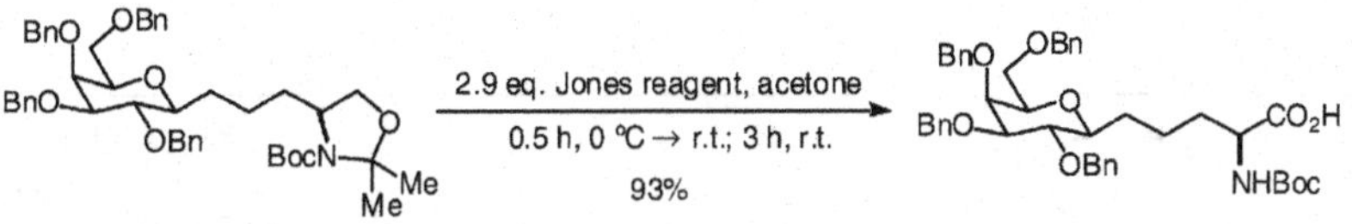

The acidic properties of Jones reagent cause hydrolysis of the oxazolidine, delivering a primary alcohol that is oxidized *in situ* to carboxylic acid. Observe the subsistence of Boc protection.[27]

Jones reagent is able to oxidize functional groups other than primary alcohols. Most notably, it is routinely employed for the transformation of secondary alcohols into ketones.[28] Molecules containing both primary and secondary alcohols can be oxidized in one step to ketoacids.[29]

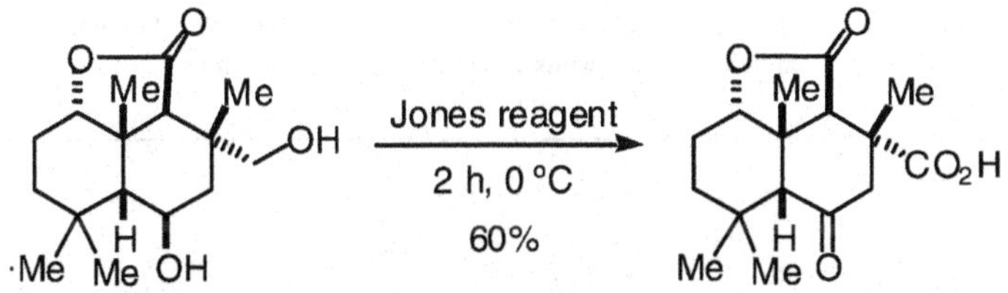

The simultaneous oxidation of a primary alcohol to a carboxylic acid and a secondary alcohol to a ketone is carried out using Jones reagent.[29a]

Zhao's Catalytic CrO₃ Oxidation

In 1998, Zhao *et al.* reported the use of catalytic CrO_3, in the presence of periodic acid (H_5IO_6) as secondary oxidant, for the transformation of primary alcohols into carboxylic acids.[30] A stock solution of reagent is prepared by dissolving H_5IO_6 (11.4 g, 50 mmol) and CrO_3 (23 mg, 1.2 mol%) in wet MeCN (0.75 v% water) to a volume of 114 mL. Carboxylic acids are normally obtained in very good yield by adding over a period of 30–60 minutes 11.4 mL of the stock solution to a solution of 2.0 mmol of alcohol in wet acetonitrile (10 mL, 0.75 v% water) kept at 0–5 °C. Under these conditions, the oxidation is normally complete in less than 30 minutes.

Although Zhao's catalytic CrO_3 oxidation has certain limitations, such as causing decomposition of electron-rich aromatics, it offers a number of advantages[16e] over Jones oxidation, including the use of a very limited amount of toxic and environmentally objectionable chromium compounds and suitability for large-scale reactions.[31]

> Zhao's research group made two important synthetic contributions to the oxidation of primary alcohols to carboxylic acids. They described both the use of catalytic CrO_3 in the presence of periodic acid as secondary oxidant[30]—Zhao's catalytic CrO_3 oxidation— and catalytic TEMPO and bleach in the presence of excess of sodium chlorite[32]—Zhao's catalytic TEMPO–bleach oxidation (see Chapter 6).

> Although Jones oxidation provides a reasonable yield of the desired diacid, Zhao's catalytic CrO_3 oxidation is preferred, because it avoids the employment of stoichiometric CrO_3 which is not practical on a large scale due to toxicity and waste disposal concerns. $RuCl_3/H_5IO_6$ delivers a low yield of diacid, probably because of destruction of electron-rich aromatics. A Swern oxidation followed by sodium chlorite ($NaClO_2$) causes epimerization of a chiral center. A TEMPO-catalyzed bleach oxidation produces significant chlorination of aromatic rings. In common with Zhao's catalytic CrO_3 oxidation, the use of sodium chlorite in the presence of catalytic TEMPO and bleach—Zhao's catalytic TEMPO/bleach oxidation—also results in good yield of the desired diacid.[31]

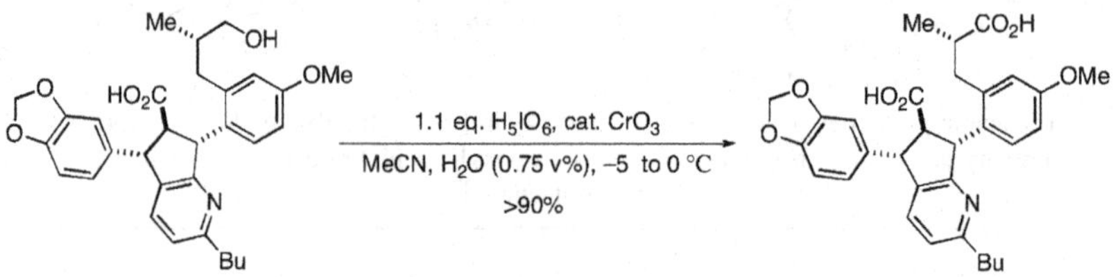

2.2. General Procedure for Oxidation of Primary Alcohols to Carboxylic Acids with Jones Reagent

A ca. 0.01–0.3 M solution of starting alcohol in acetone[a,b] is added[c] over a period of 0.5–7 hours[d] over a solution containing Jones reagent, prepared by dissolving CrO_3 (MW 100.0) in a mixture of concentrated H_2SO_4 and water.[e] Normally, from 1.0 to 10 equivalents[f,g] of CrO_3 in a ca. 1–8 M solution, prepared in a 100:58 to 100:7 (v/v) mixture of water and concentrated sulfuric acid, are used.[h,i] Once the addition is finished, the reaction mixture is left stirring until most of the starting compound is consumed.[j] The acid sometimes precipitates from the reaction mixture and can be isolated in an impure form by simple filtration. Otherwise, the excess of oxidant is normally destroyed by the addition of 2-propanol.[k] The reaction mixture is extracted with an organic solvent such as diethyl ether, ethyl acetate, or dichloromethane. The extraction can be facilitated by previous removal of the acetone by vacuum distillation. Brine is sometimes added to assist the phase separation and the extraction of acids with high solubility in water. The phase separation can be simplified by filtration of chromium salts through a pad of Celite®. Normally, the organic phase—containing the acid—is washed with water or brine. In the case of acids possessing a very adverse partition coefficient, it may be necessary to perform a continuous extraction. Finally, the acid can be isolated by two alternative protocols:

Protocol A: the organic phase is dried with Na_2SO_4 or $MgSO_4$ and concentrated, giving a crude acid that may need further purification.

Protocol B: the acid is extracted with a basic solution, such as aqueous sodium or potassium carbonate, bicarbonate or hydroxide, and then acidified to pH 2–4. The resulting aqueous phase is extracted with an organic solvent such as diethyl ether, ethyl acetate, or dichloromethane, and the collected organic phases are elaborated as in Protocol A.

[a] For best results, the alcohol must be completely dissolved in acetone. Although the oxidation sometimes proceeds satisfactorily starting from a suspension of alcohol in acetone, reduced chromium material can precipitate on the particles of suspended alcohol resulting in decreased yield of carboxylic acid.[33]

[b] The use of high concentration of alcohol in acetone may promote the generation of ester dimers.[14]

[c] The oxidation of alcohols to carboxylic acids employing Jones reagent is most commonly carried out by a "normal addition" protocol, whereby Jones reagent is added over a solution of alcohol in acetone. We are describing the so-called "inverse addition" protocol, because—although less commonly used—it normally guarantees better yield of carboxylic acid. In our opinion, the normal addition protocol must be employed only when some degree of overoxidation, due to the presence of oxidation-sensitive functional groups other than the alcohol in the molecule, is suspected.

[d] The mixing of the solutions of alcohol and oxidant is sometimes carried out over a period of only a few minutes. We recommend a much longer mixing time to avoid the formation of dimeric esters. Thus, Holland and Gilman found [14] that a mixing time as long as 7 hours may be necessary to minimize the formation of dimers. When the reaction is performed on a multigram scale, a long mixing period is necessary to avoid overheating.

[e] The oxidation must be carried out under some kind of temperature control. The reaction is most often performed at 0 °C, although temperatures as low as −25 °C are possible. The control of reaction temperature sometimes consists simply of adjusting the addition rate so as to prevent overheating above room temperature.

[f] In substrates very sensitive to overoxidation, best results are often obtained employing only 1.0 equivalent of CrO_3.

[g] Sometimes, the quantity of oxidant is adjusted using a titrating protocol in which a "normal" addition is employed and Jones reagent is added to the solution of alcohol in acetone until persistence of Jones reagent color is observed. This may lead to incomplete oxidation, because the consumption of Jones reagent during the oxidation is not instantaneous. [34]

[h] Jones reagent is very often prepared as detailed by Djerassi et al. [15] by diluting a solution of 26.72 g of CrO_3 in 23 mL of concentrated sulfuric acid with water to a volume of 100 mL.

[i] Higher concentration of sulfuric acid leads to Jones reagent possessing greater oxidizing power. On the other hand, this may promote side reactions on acid-sensitive functional groups.

[j] It normally takes from 20 minutes to 6 hours.

[k] The quenching of Jones reagent is readily noticed by a conspicuous color change from reddish to green. The reaction is sometimes quenched with methanol.

In this challenging substrate, best conditions for obtention of acid were met with Jones oxidation. According to the authors: "The presence of a reactive double bond and its propensity to isomerize from the β, γ-position to the α, β-unsaturated position precluded the use of many oxidation procedures such as $RuCl_3/NaIO_4$, $KMnO_4/AcOH$, silver oxide, TEMPO radical-based methodologies and sodium bromite. In most cases, degradation of the substrate was observed or α, β-dehydroamino acids or the starting alcohol was isolated. To prevent scrambling of the easily racemizable chiral center, oxidations under basic conditions had to be avoided (e.g. $KMnO_4/NaOH$)." [35]

A Jones oxidation on a multigram scale is described. [36]

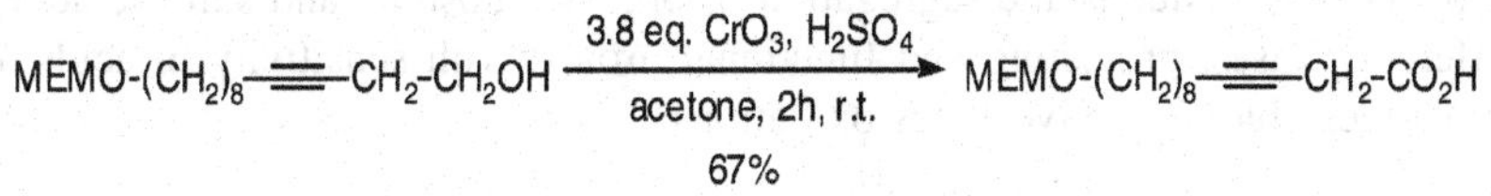

$$\text{MEMO-(CH}_2)_8\text{---}\equiv\text{---CH}_2\text{-CH}_2\text{OH} \xrightarrow[\substack{\text{acetone, 2h, r.t.} \\ 67\%}]{\text{3.8 eq. CrO}_3,\ \text{H}_2\text{SO}_4} \text{MEMO-(CH}_2)_8\text{---}\equiv\text{---CH}_2\text{-CO}_2\text{H}$$

While oxidation with PDC in DMF provides low yield of the desired β, γ-unsaturated carboxylic acid, the use of Jones oxidation with the "inverse addition" protocol at low temperature allows the obtention of a reasonable yield of the desired acid.[16b]

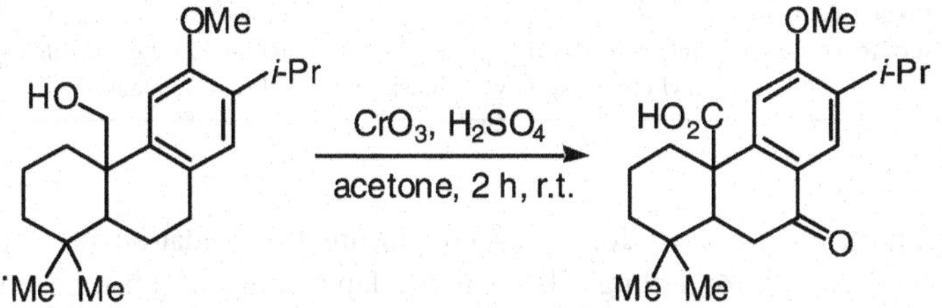

This hindered alcohol is not oxidized with hydrated copper permanganate in CH$_2$Cl$_2$, solid sodium permanganate, or potassium ruthenate. It is possible to achieve the desired oxidation to acid employing Jones reagent at room temperature, resulting in an additional oxidation at the benzylic position. Increasing the reaction time leads to a mixture of compounds.[37]

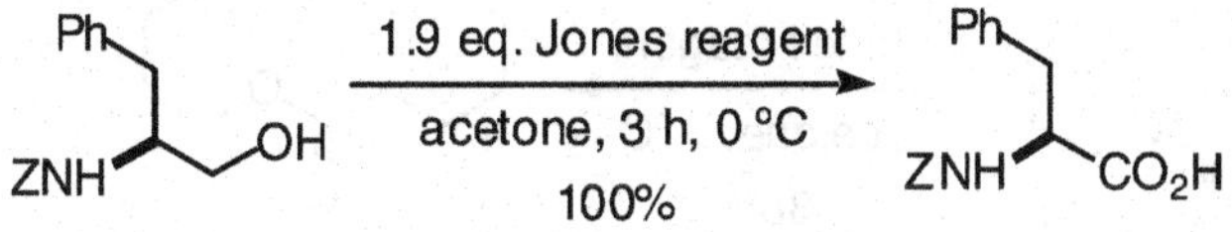

Jones reagent was chosen for this oxidation because it produced very little racemization. Alkaline potassium permanganate was not effective because of lack of solubility of the starting alcohol in aqueous sodium hydroxide. A two-step oxidation via the corresponding aldehyde was discarded because CrO$_3$/Py and DCC/DMSO yielded a substantially racemized aldehyde.[38]

2.3. Functional Group and Protecting Group Sensitivity to Jones Oxidation

Information regarding functional group and protecting group sensitivity to Jones oxidation, when it is employed in the transformation of secondary alcohols into ketones, can be found in the first volume of this series.[39]

Functional groups with some sensitivity to oxidative or acidic conditions are sometimes affected by Jones oxidation during the obtention of carboxylic acids. However, due to the segregation of organic substrate and sulfuric acid in different phases, protecting and functional groups with sensitivity to acids are sometimes able to survive Jones oxidations.

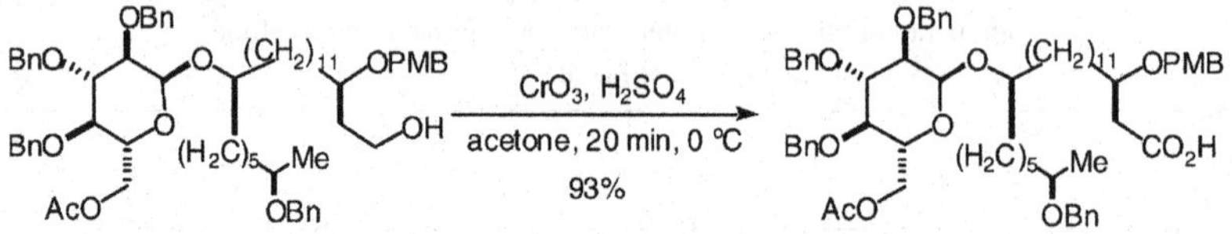

A primary alcohol is successfully oxidized to carboxylic acid with Jones reagent without interference with a number of sensitive protecting and functional groups, including a p-methoxybenzyl ether that is very sensitive to acids and oxidants.[40]

TBDPS ethers resist Jones reagent during the oxidation of alcohols to carboxylic acids.[41] Regarding TBS ethers, depending on substrate and exact reaction conditions, it is possible to oxidize alcohols to carboxylic acids in the presence of unreacting TBS ethers[42]; also, it is possible to perform the hydrolysis of a TBS ether and the *in situ* oxidation of the resulting alcohol to carboxylic acid.[42b, 43]

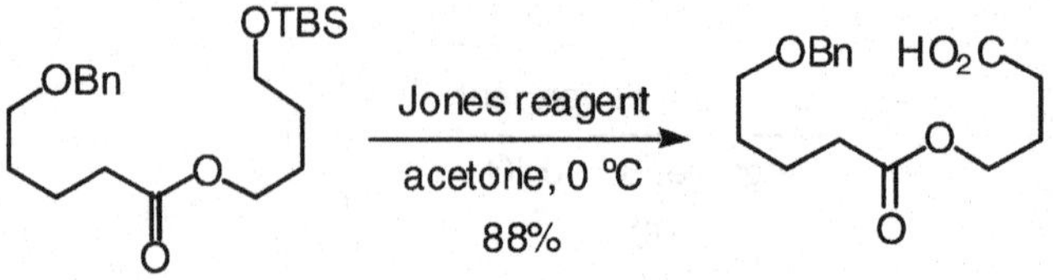

A TBS ether is hydrolyzed under Jones conditions and the resulting alcohol is oxidized *in situ* to carboxylic acid.[42b]

Both the robust benzyl ethers[40, 44] and the less robust p-methoxybenzyl ethers,[40, 45] which possess more sensitivity to oxidative and acidic conditions, are able to withstand the action of Jones reagent.

Benzyl ethers are known to be oxidized under the action of Jones reagent under immoderate conditions.[46] As the oxidation of alcohols requires much milder conditions, benzyl ethers do not normally interfere with the oxidation of alcohols to carboxylic acids.

Both MOM[47] and benzyloxymethyl[48] ethers resist Jones oxidation, while the more acid-sensitive THP ethers are hydrolyzed and the resulting alcohols are transformed into carboxylic acids.[49]

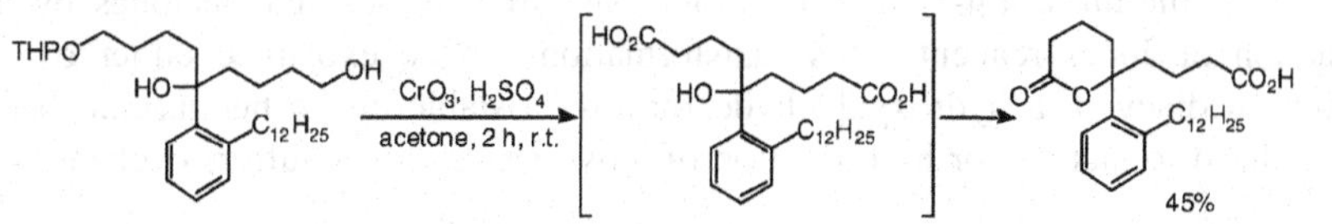

The very acid-sensitive THP ether is hydrolyzed under the acidic conditions of Jones oxidation, resulting in an alcohol that is transformed *in situ* into a carboxylic acid that can be isolated as lactone.[49]

Regarding acetals protecting 1,2- and 1,3-diols, it has been reported that benzylidene,[50] isopropylidene,[51] cyclohexylidene,[19] and methylene[52] acetals resist the action of Jones reagent while an alcohol is oxidized to a carboxylic acid.

Depending on substrate and exact reaction conditions, dioxolanes protecting aldehydes and ketones can both be hydrolyzed[53] or remain unchanged[54] during the oxidation of primary alcohols with Jones reagent.

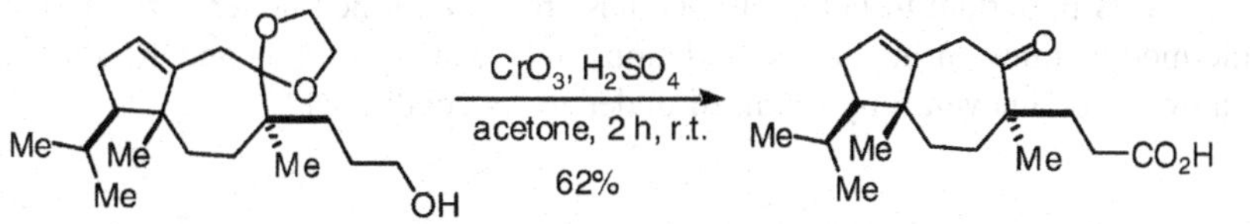

A dioxolane protecting group is removed while an alcohol is oxidized to an acid with Jones reagent. Observe that acid-catalyzed migration of the alkene into conjugation with the ketone does not occur.[53b]

Most amine protecting groups resist the action of Jones reagent. These include the ubiquitous acetyl, benzyl, Fmoc,[55] Z,[44, 48, 56] and even the acid-sensitive Boc[47d, 57] groups.

Free secondary alcohols are normally oxidized to ketones during the treatment of primary alcohols with Jones reagent.[52a, 58] Nevertheless, it is possible to oxidize primary alcohols in the presence of unreacting secondary ones, resulting in formation of hydroxyacids in yields fluctuating from modest to fair.[59]

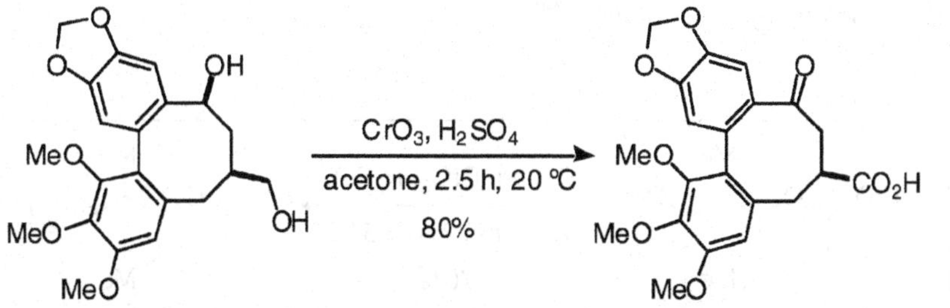

Jones reagent performs the simultaneous oxidation of a primary alcohol to carboxylic acid and a secondary alcohol to ketone.[52a]

Sometimes, 1,4-[60] and 1,5-diols[61] are transformed into lactones by the action of Jones reagent. This transformation can be explained either by the intermediacy of a hydroxyaldehyde forming a stable cyclic hemiacetal that is oxidized to lactone, or by formation of a hydroxyacid that suffers cyclization.

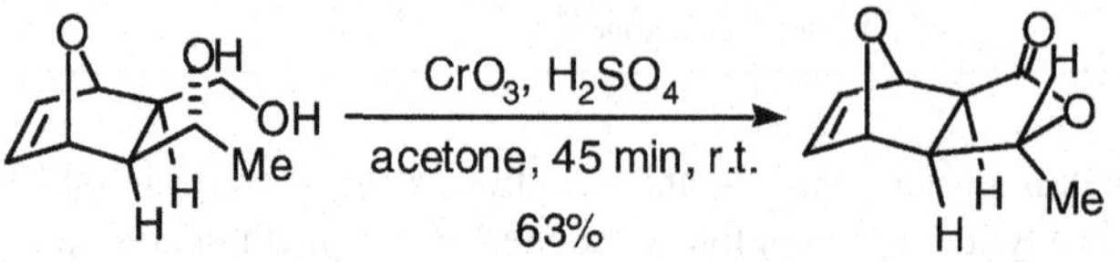

During the oxidation of a primary alcohol with Jones reagent, a secondary alcohol, rather than being transformed into a ketone, forms a stable lactone.[60a]

It is important to note that lactones are only formed under very favorable thermodynamics during Jones oxidation, and a variety of 1,4- and 1,5-diols are known to be uneventfully oxidized under Jones conditions.[62]

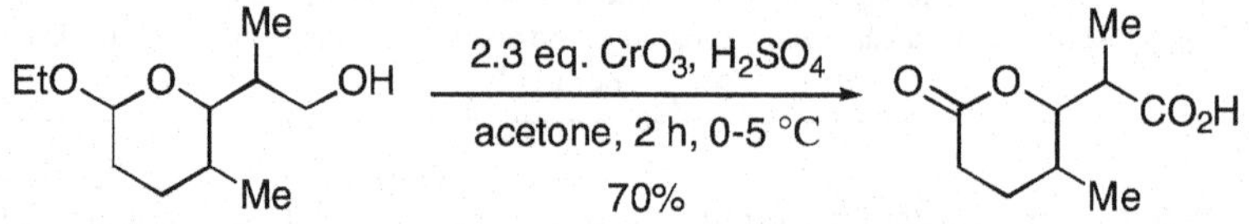

A 1,4-diol is uneventfully oxidized to a ketoacid with Jones reagent, with no formation of a lactone, which otherwise would look stable, being reported.[62c]

Acetals, belonging to diverse structural types, normally resist the action of Jones reagent. Examples include glycosides[40, 44, 63] and spiroacetals.[41a,b,e] Nonetheless, some cases are reported in which a cyclic acetal is transformed into a lactone by Jones reagent.[64]

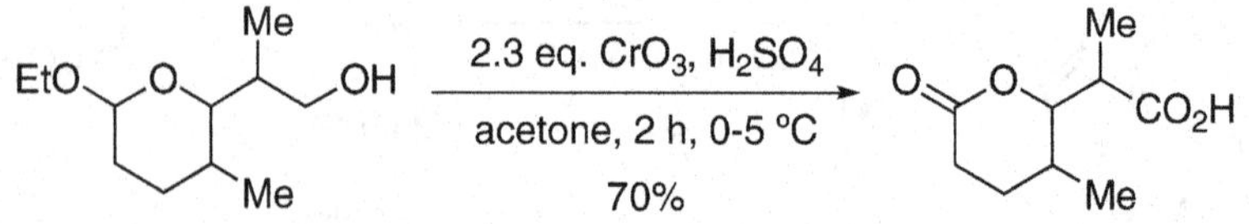

This is a rare case in which an acetal is oxidized during the transformation of a primary alcohol into carboxylic acid with Jones reagent.[64]

Aldehydes are very quickly oxidized with Jones reagent. In fact, it is possible to perform the selective oxidation of an aldehyde in the presence of free alcohols.[65] That is why intermediate aldehydes are very rarely isolated during the obtention of carboxylic acids with Jones reagent. Quite expectedly, by using excess of Jones reagent, it is possible to oxidize both aldehydes and primary alcohols to carboxylic acids.[66]

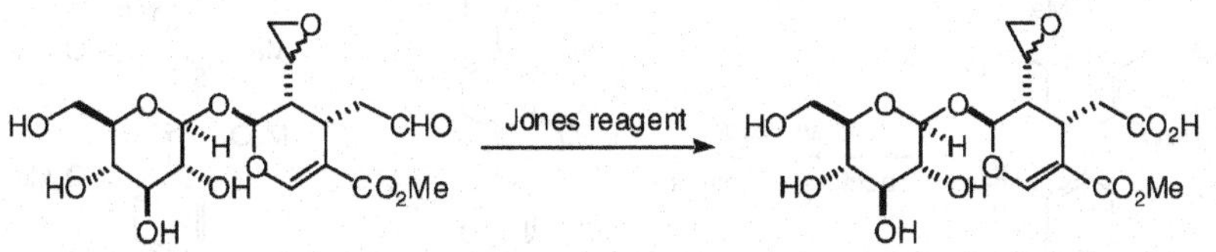

An aldehyde is oxidized to a carboxylic acid with Jones reagent while one primary alcohol and three secondary alcohols remain unchanged.[65]

Epoxides normally resist Jones oxidation during the preparation of carboxylic acids.[67]

Amines, including the quite oxidation-prone primary amines,[68] normally resist[21, 47c, 69] Jones reagent during the preparation of carboxylic acids. This happens probably due to protection of the amines by protonation. This result is quite useful, because very few oxidants are able to preserve primary amines intact during generation of carboxylic acids.

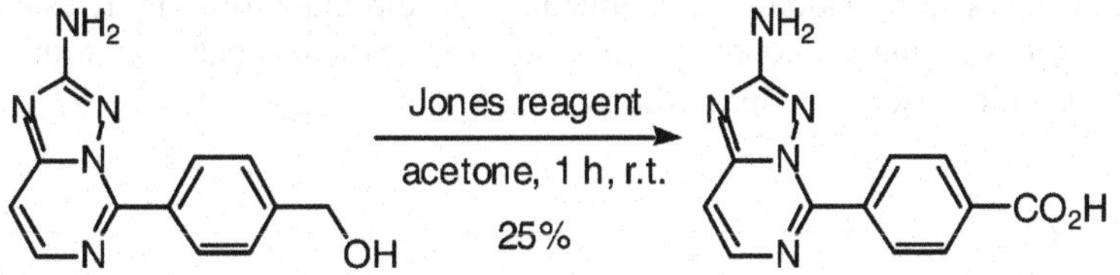

An alcohol, in a substrate possessing a primary amine with a high sensitivity to oxidation, is transformed into the corresponding carboxylic acid in moderate yield. The amine remains unchanged possibly because of protection by protonation.[68]

2.4. Side Reactions

Information regarding side reactions during the oxidation of secondary alcohols with Jones reagent can be found in the volume of this series entitled *Oxidation of Alcohols to Aldehydes and Ketones.*[70]

Primary alcohols occasionally suffer an incomplete oxidation with Jones reagent, resulting in formation of aldehydes.[71] As aldehydes are transformed into carboxylic acids with Jones reagent via the corresponding hydrates, aldehydes possessing a low proportion of hydrate in equilibrium are more likely to remain unchanged. Aldehydes with a low proportion of hydrate include hindered aldehydes and aldehydes conjugated with alkenes or aromatic rings.[9]

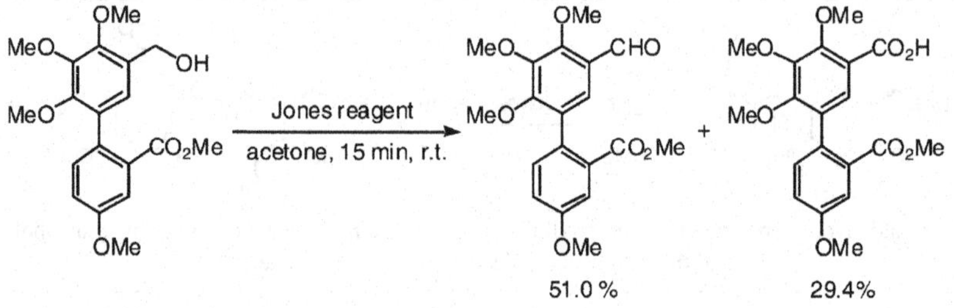

51.0%   29.4%

A 51% yield of aldehyde is isolated from the oxidation of an alcohol with Jones reagent. The following factors favor an incomplete oxidation of the alcohol: (1) short reaction time, (2) an intermediate aldehyde possessing steric hindrance that disfavors formation of hydrate, and (3) the presence of electron-donating methoxy groups on the benzene ring preventing the buildup of a high concentration of aldehyde hydrate.[71b]

Sometimes, dimeric esters are obtained from the oxidation of primary alcohols with Jones reagent (see page 15).[11] These compounds are formed from the oxidation of intermediate hemiacetals, whose generation is favored by the buildup of a high concentration of starting alcohol and intermediate aldehyde. As demonstrated by Holland and Gilman,[14] the formation of dimeric esters can be mitigated by the use of an "inverse addition" protocol rather than the more common "direct addition" procedure.

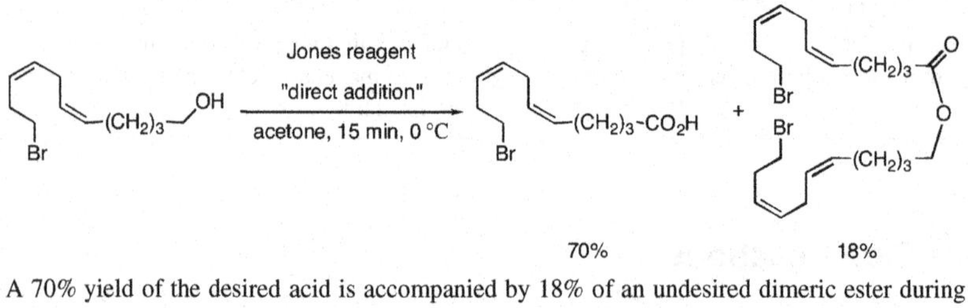

70%   18%

A 70% yield of the desired acid is accompanied by 18% of an undesired dimeric ester during the oxidation of an alcohol with Jones reagent employing a "direct addition" protocol. Probably, the use of the "inverse addition" protocol recommended by Holland and Gilman would lead to a decreased amount of dimeric ester.[11b]

During the oxidation of primary alcohols with Jones reagent, the intermediate chromate ester normally evolves as in path **a** below, leading to an aldehyde that is further oxidized to acid. On the other hand, sometimes, the intermediate chromate ester evolves as in path **b**, resulting in a carbon–carbon bond breakage in which formaldehyde and a carbocation R^+ are formed.[72] This side reaction is facilitated by the stability of carbocation R^+.

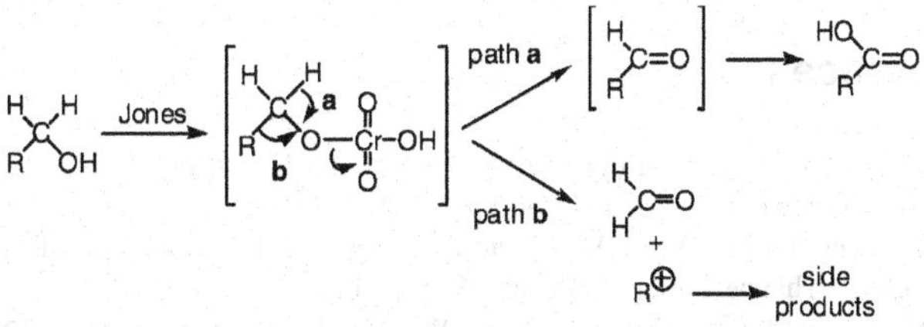

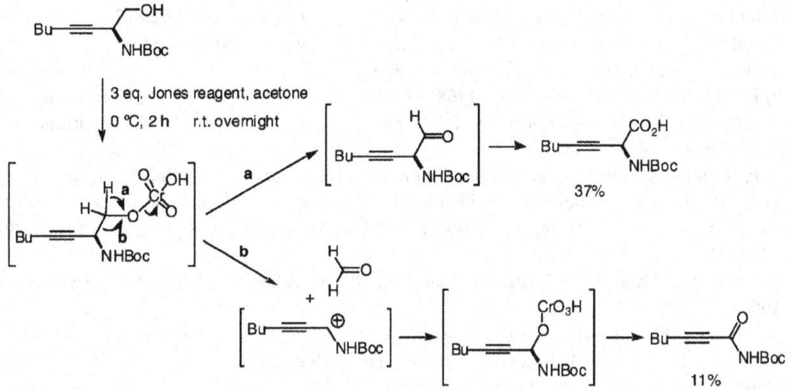

The treatment of an alcohol with Jones reagent leads to 37% of the desired carboxylic acid, accompanied by 11% of an *N*-Boc amide, formed from an intermediate carbocation stabilized by a nitrogen atom and by conjugation with an alkyne.[72b]

It is important to note that a carbon–carbon bond breakage during the oxidation of a primary alcohol with Jones reagent depends not only on the stability of the resulting carbocation, but also on a proper orbital alignment in the preceding chromate ester, the likeness of which is difficult to anticipate. Therefore, an attempt to perform a Jones oxidation to carboxylic acid must not be avoided due to the sole concern for this side reaction.

Sometimes, because of the acidic nature of Jones reagent, acid-catalyzed migrations of alkenes facilitated by favorable thermodynamics are observed during the oxidation of alcohols to carboxylic acids.[73]

$$HO_2C-C\equiv C-CH_2-CH_2OH \xrightarrow[\text{acetone, 20 °C, 2 h}]{\text{Jones reagent}} HO_2C-CH=C=CH-CO_2H$$

23%

An alkyne is isomerized to a highly conjugated allene during the oxidation of an alcohol with Jones reagent.[73]

2.5. References

1 a) Poos, G. I.; Arth, G. E.; Beyler, R. E.; Sarett, L. H.; *J.Am.Chem.Soc.* **1953**, *75*, 422. b) Zibuck, R.; Streiber, J.; *Org.Synth.Coll.Vol.IX* **1998**, 432.

2 Bosche, H. G. in Houben-Weyl, *Methoden der organischen Chemie*, 4th ed.; E. Müller, Ed., Vol. 4/1b; Georg Thieme Verlag: Stuttgart, 1975, p. 429.

3 Bowden, K.; Heilbron, I. M.; Jones, E. R. H.; Weedon, B. C. L.; *J.Chem.Soc.* **1946**, 39.

4 a) Heilbron, I.; Jones, E. R. H.; Sondheimer, F.; *J.Chem.Soc.* **1947**, 1586. b) Heilbron, I.; Jones, E. R. H.; Sondheimer, F.; *J.Chem.Soc.* **1949**, 604.

5 a) Lanes, R. M.; Lee, D. G.; *J.Chem.Edoc.* **1968**, *45*, 269. b) Westheimer, F. H.; Nicolaides, N.; *J.Am.Chem.Soc.* **1949**, *71*, 25. c) Roček, J.; Ng, C.-S.; *J.Org.Chem.* **1973**, *38*, 3348.

6 Vogel, A. I. *Practical Organic Chemistry*, 3rd ed.; Longman: London, p. 318.

7 Veliev, M. G.; Guseinov, M. M.; *Synthesis* **1980**, 461.

8 a) Le Hénaff, P.; *Bull.Soc.Chim.Fr.* **1968**, *11*, 4689. b) Lindsay Smith, J. R.; Norman, R. O. C.; Stillings, M. R.; *Tetrahedron* **1978**, *34*, 1381. c) Durst, T.; Kozma, E. C.; Charlton, J. L.; *J.Org.Chem.* **1985**, *50*, 4829.

9 a) F. Barrero, A.; Altarejos, J.; Álvarez-Manzaneda, E. J.; Ramos, J. M.; Salido, S.; *Tetrahedron* **1993**, *49*, 6251. b) Fringuelli, F.; Pellegrino, R.; Piermatti, O.; Pizzo, F.; *Synth.Commun.* **1994**, *24*, 2665. c) Sun, M.; Deng, Y.; Batyreva, E.; Sha, W.; Salomon, R. G.; *J.Org.Chem.* **2002**, *67*, 3575.

10 Shishido, K.; Hiroya, K.; Komatsu, H.; Fukumoto, K.; Kametani, T.; *J.Chem.Soc.Chem.Commun.* **1986**, *12*, 904.

11 a) Marples, B. A.; Saint, C. G.; Traynor, J. R.; *J.Chem.Soc. Perkin Trans. I* **1986**, 567. b) Mosset, P.; Grée, R.; Falck, J. R.; *Synth.Commun.* **1989**, *19*, 645.

12 Craig, J. C.; Horning, E. C.; *J.Org.Chem.* **1960**, *25*, 2098.

13 Eglinton, G.; Whiting, M. C.; *J.Chem.Soc.* **1953**, 3052.

14 Holland, B. C.; Gilman, N. W.; *Synth.Commun.* **1974**, *4*, 203.

15 Djerassi, C.; Engle, R. R.; Bowers, A.; *J.Org.Chem.* **1956**, *21*, 1547.

16 See, for example: a) Grieco, P. A.; *J.Am.Chem.Soc.* **1969**, *91*, 5660. b) Millar, J. G.; Oehlschlager, A. C.; Wong, J. W.; *J.Org.Chem.* **1983**, *48*, 4404. c) Manfre, F.; Marc Kern, J.; Biellmann, J. F.; *J.Org.Chem.* **1992**, *57*, 2060. d) Reginato, G.; Mordini, A.; Valacchi, M.; *Tetrahedron Lett.* **1998**, *39*, 9545. e) Reginato, G.; Mordini, A.; Valacchi, M.; Grandini, E.; *J.Org.Chem.* **1999**, *64*, 9211.

17 Grob, C. A.; Wittwer, G.; Rao, K. R.; *Helv.Chim.Acta* **1985**, *68*, 760.

18 Sita, L. R.; *J.Org.Chem.* **1993**, *58*, 5285.

19 Williams, D. R.; Benbow, J. W.; *J.Org.Chem.* **1988**, *53*, 4643.

20 Fuentes, L. M.; Shinkai, I.; King, A.; Purick, R.; Reamer, R. A.; Schmitt, S. M.; Cama, L.; Christensen, B. G.; *J.Org.Chem.* **1987**, *52*, 2563.

21 Clark, J. S.; Hodgson, P. B.; Goldsmith, M. D.; Blake, A. J.; Cooke, P. A.; Street, L. J.; *J.Chem.Soc. Perkin Trans. I* **2001**, *24*, 3325.

22 a) Pattison, F. L. M.; *J.Am.Chem.Soc.* **1956**, *78*, 2255. b) Evans, A. R.; Martin, R.; Taylor, G. A.; Yap, C. H. M.; *J.Chem.Soc. Perkin Trans. I* **1987**, *7*, 1635.

23 a) Olsson, R. A.; Kusachi, S.; Thompson, R. D.; Ukena, D.; Padgett, W.; Daly, J. W.; *J.Med.Chem.* **1986,** *29,* 1683. b) Rasmusson, G. H.; Reynolds, G. F.; Steinberg, N. G.; Walton, E.; Patel, G. F.; Liang, T.; Cascieri, M. A.; Cheung, A. H.; Brooks, J. R.; Berman, C.; *J.Med.Chem.* **1986,** *29,* 2298. c) Terunuma, D.; Motegi, M.; Tsuda, M.; Sawada, T.; Nozawa, H.; Nohira, H.; *J.Org.Chem.* **1987,** *52,* 1630. d) Wittenberg, D.; Talukdar, P. B.; Gilman, H.; *J.Am.Chem.Soc.* **1960,** *82,* 3608. e) Schmidt, R. R.; Fritz, H.-J.; *Chem.Ber.* **1970,** *103,* 1867. f) Sá, M. M.; Silveira, G. P.; Castilho, M. S.; Pavão, F.; Oliva, G.; *ARKIVOC* **2002,** *VIII,* 112. g) Avery, M. A.; Fan, P.; Karle, J. M.; Miller, R.; Goins, K.; *Tetrahedron Lett.* **1995,** *36,* 3965. h) Avery, M. A.; Fan, P.; Karle, J. M.; Bonk, J. D.; Miller, R.; Goins, D. K.; *J.Med.Chem.* **1996,** *39,* 1885.

24 Cornforth, R. H.; Cornforth, J. W.; Popják, G.; *Tetrahedron* **1962,** *18,* 1351.

25 a) Newman, M. S.; Arkell, A.; Fukunaga, T.; *J.Am.Chem.Soc.* **1960,** *82,* 2498. b) Mathur, H. H.; Bhattacharyya, S. C.; *J.Chem.Soc.* **1963,** 3505.

26 Katajisto, J.; Karskela, T.; Heinonen, P.; Lönnberg, H.; *J.Org.Chem.* **2002,** *67,* 7995.

27 Dondoni, A.; Mariotti, G.; Marra, A.; *J.Org.Chem.* **2002,** *67,* 4475.

28 Tojo, G.; Fernández, M. in *Basic Reactions in Organic Synthesis. Oxidation of Alcohols to Aldehydes and Ketones: A Guide to Current Common Practice,* G. Tojo, Ed.; Springer: New York, 2006, p. 5.

29 See for example: a) Ziegler, F. E.; Jaynes, B. H.; Saindane, M. T.; *Tetrahedron Lett.* **1985,** *26,* 3307. b) Kim, D. S. H. L.; Chen, Z.; Nguyen, van T.; Pezzuto, J. M.; Qiu, S.; Lu, Z.-Z.; *Synth.Commun.* **1997,** *27,* 1607. c) Noguchi, T.; Onodera, K.; Tomisawa, K.; Yokomori, S.; *Chem.Pharm.Bull.* **2002,** *50,* 1407.

30 Zhao, M.; Li, J.; Song, Z.; Desmond, R.; Tschaen, D. M.; Grabowski, E. J. J.; Reider, P. J.; *Tetrahedron Lett.* **1998,** *39,* 5323.

31 Song, Z. J.; Zhao, M.; Desmond, R.; Devine, P.; Tschaen, D. M.; Tillyer, R.; Frey, L.; Heid, R.; Xu, F.; Foster, B.; Li, J.; Reamer, R.; Volante, R.; Grabowski, E. J. J.; Dolling, U. H.; Reider, P. J.; Okada, S.; Kato, Y.; Mano, E.; *J.Org.Chem.* **1999,** *64,* 9658.

32 Zhao, M.; Li, J.; Mano, E.; Song, Z.; Tschaen, D. M.; Grabowski, E. J. J.; Reider, P. J.; *J.Org.Chem.* **1999,** *64,* 2564.

33 Robert, N.; Cadiot, M. P.; *Bull.Soc.Chim.Fr.* **1956,** *93,* 1579.

34 See, for example: Wolinsky, J.; Gibson, T.; Chan, D.; Wolf, H.; *Tetrahedron* **1965,** *21,* 1247.

35 Beaulieu, P. L.; Duceppe, J.-S.; Johnson, C.; *J.Org.Chem.* **1991,** *56,* 4196.

36 Acheson, R. M.; Verlander, M. S.; *J.Chem.Soc. (C)* **1969,** 2311.

37 Banerjee, A. K.; Hurtado, S. H.; Laya, M. M.; Acevedo, J. C.; Álvarez, G. J.; *J.Chem.Soc. Perkin Trans. I* **1988,** 931.

38 Kashima, C.; Harada, K.; Fujioka, Y.; Maruyama, T.; Omote, Y.; *J.Chem.Soc. Perkin Trans. I* **1988,** 535.

39 Tojo, G.; Fernández, M. in *Basic Reactions in Organic Synthesis. Oxidation of Alcohols to Aldehydes and Ketones: A Guide to Current Common Practice,* G. Tojo, Ed.; Springer: New York, 2006, p. 8–9.

40 Takahashi, S.; Souma, K.; Hashimoto, R.; Koshino, H.; Nakata, T.; *J.Org.Chem.* **2004,** *69,* 4509.

41 a) Boeckman Jr., R. K.; Charette, A. B.; Asberom, T.; Johnston, B. H.; *J.Am.Chem.Soc.* **1991,** *113,* 5337. b) Boeckman Jr., R. K.; Charette, A. B.; Asberom, T.; Johnston, B. H.; *J.Am.Chem.Soc.* **1987,** *109,* 7553. c) Hillis, L. R.; Ronald, R. C.; *J.Org.Chem.* **1985,** *50,* 470. d) Berlage, U.; Schmidt, J.; Peters, U.; Welzel, P.; *Tetrahedron Lett.* **1987,** *28,* 3091. e) Nakahara, Y.; Fujita, A.; Beppu, K.; Ogawa, T.; *Tetrahedron* **1986,** *42,* 6465.

42 a) Marshall, J. A.; Audia, J. E.; Shearer, B. G.; *J.Org.Chem.* **1986,** *51,* 1730. b) Nicolaou, K. C.; Hwang, C. K.; Marron, B. E.; DeFrees, S. A.; Couladouros, E. A.; Abe, Y.; Carroll, P. J.; Snyder, J. P.; *J.Am.Chem.Soc.* **1990,** *112,* 3040.

43 a) Young, R. N.; Champion, E.; Gauthier, J. Y.; Jones, T. R.; Leger, S.; Zamboni, R.; *Tetrahedron Lett.* **1986,** *27,* 539. b) Nakazato, A.; Kumagai, T.; Sakagami, K.; Yoshikawa, R.; Suzuki, Y.; Chaki, S.; Ito, H.; Taguchi, T.; Nakanishi, S.; Okuyama, S.; *J.Med.Chem.* **2000,** *43,* 4893.

44 Beetz, T.; van Boeckel, C. A. A.; *Tetrahedron Lett.* **1986,** *27,* 5889.

45 Horita, K.; Oikawa, Y.; Nagato, S.; Yonemitsu, O.; *Tetrahedron Lett.* **1988**, *29*, 5143.

46 Bal, B. S.; Kochhar, K. S.; Pinnick, H. W.; *J.Org.Chem.* **1981**, *46*, 1492.

47 a) Marshall, J. A.; Audia, J. E.; Grote, J.; Shearer, B. G.; *Tetrahedron***1986**, *42*, 2893.
    b) Plata, D. J.; Kallmerten, J.; *J.Am.Chem.Soc.* **1988**, *110*, 4041. c) Fukuyama, T.; Nunes, J. J.;
    J.Am.Chem.Soc. **1988**, *110*, 5196. d) Jourdant, A.; Zhu, J.; *Tetrahedron Lett.* **2000**, *41*, 7033.

48 Seki, M.; Shimizu, T.; Inubushi, K.; *Synthesis* **2002**, *3*, 361.

49 Perchonock, C. D.; Uzinskas, I.; McCarthy, M. E.; Erhard, K. F.; Gleason, J. G.; Wasserman,
    M. A.; Muccitelli, R. M.; DeVan, J. F.; Tucker, S. S.; Vickery, L. M.; Kirchner, T.; Weichman,
    B. M.; Mong, S.; Scott, M. O.; Chi-Rosso, G.; Wu, H.-L.; Crooke, S. T.; Newton, J. F.;
    J.Med.Chem. **1986**, *29*, 1442.

50 a) Madaj, J.; Jankowska, M.; Winiewski, A.; *Carbohydr.Res.* **2004**, *339*, 1293. b) Hamada, T.;
    Kiyokawa, M.; Kobayashi, Y.; Yoshioka, T.; Sano, J.; Yonemitsu, O.; *Tetrahedron* **2004**,
    60, 4693.

51 a) Ho Kang, S.; Jun, H.-S.; *Chem.Commun.* **1998**, *18*, 1929. b) Ghosh, A. K.; Bilcer, G.; *Tetrahedron
    Lett.* **2000**, *41*, 1003. c) Adachi, H.; Nishimura, Y.; Takeuchi, T.; *J.Antibiot.* **2002**, *55*, 92.

52 a) Magnus, P.; Schultz, J.; Gallagher, T.; *J.Am.Chem.Soc.* **1985**, *107*, 4984. b) Lima, P. C.; Lima,
    L. M.; da Silva, K. C. M.; Léda, P. H. O.; de Miranda, A. L. P.; Fraga, C. A. M.; Barreiro, E. J.;
    Eur.J.Med.Chem. **2000**, *35*, 187.

53 a) Toyota, M.; Odashima, T.; Wada, T.; Ihara, M.; *J.Am.Chem.Soc.* **2000**, *122*, 9036. b) Tan,
    D. S.; Dudley, G. B.; Danishefsky, S. J.; *Angew.Chem.Int.Ed.* **2002**, *41*, 2185.

54 Horita, K.; Oikawa, Y.; Yonemitsu, O.; *Chem.Pharm.Bull.* **1989**, *37*, 1698.

55 a) Cini, N.; Machetti, F.; Menchi, G.; Occhiato, E. G.; Guarna, A.; *Eur.J.Org.Chem.* **2002**, *5*,
    873. b) Virta, P.; Rosenberg, J.; Karskela, T.; Heinonen, P.; Lönnberg, H.; *Eur.J.Org.Chem.*
    2001, *18*, 3467. c) Ohnishi, Y.; Ichikawa, Y.; *Bioorg.Med.Chem.Lett.* **2002**, *12*, 997.

56 Li, W.; Hanan, C. E.; d'Avignon, A.; Moeller, K. D.; *J.Org.Chem.* **1995**, *60*, 8155.

57 a) Sagnard, I.; Sasaki, A.; Chiaroni, A.; Riche, C.; Potier, P.; *Tetrahedron Lett.* **1995**, *36*, 3149.
    b) Kirihara, M.; Kawasaki, M.; Takuwa, T.; Kakuda, H.; Wakikawa, T.; Takeuchi, Y.; Kirk,
    K. L.; *Tetrahedron: Asymmetry* **2003**, *14*, 1753.

58 Hashimoto, S.; Arai, Y.; Hamanaka, N.; *Tetrahedron Lett.* **1985**, *26*, 2679.

59 Allanson, N.M.; Liu, D.; Feng, C.; Jain, R.K.; Chen, A.; Ghosh, M.; Hong, L.; Sofia, M.J.;
    Tetrahedron Lett. **1998**, *39*, 1889.

60 a) Bloch, R.; Gilbert, L.; *J.Org.Chem.* **1987**, *52*, 4603. b) Weyerstahl, P.; Marschall-
    Weyerstahl, H.; Penninger, J.; Walther, L.; *Tetrahedron* **1987**, *43*, 5287. c) Blay, G.; Cardona, L.;
    García, B.; Pedro, J. R.; *J.Nat.Prod.* **1993**, *56*, 1723.

61 Lochynski, S.; Jarosz, B.; Walkowicz, M.; Piatkowski, K.; *J.Prakt.Chem.* **1988**, *330*, 284.

62 a) Randad, R. S.; Kulkarni, G. H.; *Ind.J.Chem.* **1985**, *24B*, 225. b) Kogen, H.; Tomioka, K.;
    Hashimoto, S.-I.; Koga, K.; *Tetrahedron* **1981**, *37*, 3951. c) Chenault, H. K.; Yang, J.; Taber,
    D. F.; *Tetrahedron* **2000**, *56*, 3673.

63 Wessel, H. P.; Labler, L.; Tschopp, T. B.; *Helv.Chim.Acta* **1989**, *72*, 1268.

64 Malanga, C.; Menicagli, R.; Dell'Innocenti, M.; Lardicci, L.; *Tetrahedron Lett.* **1987**, *28*, 239.

65 Kuwajima, H.; Tanahashi, T.; Inoue, K.; Inouye, H.; *Chem.Pharm.Bull.* **1999**, *47*, 1634.

66 Känel, H.-R.; Ganter, C.; *Helv.Chim.Acta* **1985**, *68*, 1226.

67 Tius, M. A.; Trehan, S.; *J.Org.Chem.* **1989**, *54*, 46.

68 Medwid, J. B.; Paul, R.; Baker, J. S.; Brockman, J. A.; Du, M. T.; Hallett, W. A.; Hanifin, J. W.;
    Hardy Jr., R. A.; Tarrant, M. E.; Torley, L. W.; Wrenn, S.; *J.Med.Chem.* **1990**, *33*, 1230.

69 a) Baker, B. R.; Schaub, R. E.; Williams, J. H.; *J.Org.Chem.* **1952**, *17*, 109. b) Yong, Y. F.;
    Lipton, M. A.; *Bioorg.Med.Chem.Lett.* **1993**, *3*, 2879. c) Smith, A. L.; Williams, S. F.; Holmes,
    A. B.; Hughes, L. R.; Lidert, Z.; Swithenbank, C.; *J.Am.Chem.Soc.* **1988**, *110*, 8696. d) Tong,
    M. K.; Blumenthal, E. M.; Ganem, B.; *Tetrahedron Lett.* **1990**, *31*, 1683.

70 Tojo, G.; Fernández, M. in *Basic Reactions in Organic Synthesis. Oxidation of Alcohols to
    Aldehydes and Ketones: A Guide to Current Common Practice*, G. Tojo, Ed.; Springer: New
    York, 2006, p. 12.

71 a) Dupont, G.; Dulou, R.; Lefebvre, G.; *Bull.Soc.Chim.Fr.* **1951,** 339. b) Itoh, Y.; Brossi, A.;
     Hamel, E.; Flippen-Anderson, J. L.; George, C.; *Helv.Chim.Acta* **1989,** *72,* 196.

72 a) He, J.-F.; Wu, Y.-L.; *Tetrahedron* **1988,** *44,* 1933. b) Meffre, P.; Gauzy, L.; Branquet, E.;
     Durand, P.; Le Goffic, F.; *Tetrahedron* **1996,** *52,* 11215. c) Dondoni, A.; Massi, A.; Minghini, E.;
     Sabbatini, S.; Bertolasi, V.; *J.Org.Chem.* **2003,** *68,* 6172.

73 Jones, E. R. H.; Mansfield, G. H.; Whiting, M. C.; *J.Chem.Soc.* **1954,** 3208.

3

Pyridinium Dichromate (PDC) in Dimethylformamide. The Method of Corey and Schmidt

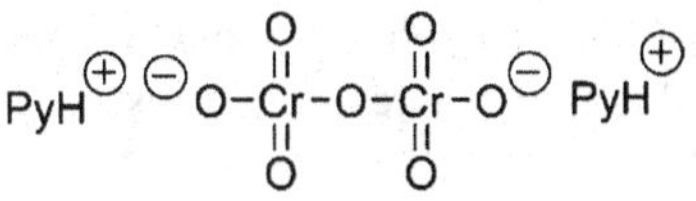

$$PyH^{\oplus}\ ^{\ominus}O-\underset{\underset{O}{\parallel}}{\overset{\overset{O}{\parallel}}{Cr}}-O-\underset{\underset{O}{\parallel}}{\overset{\overset{O}{\parallel}}{Cr}}-O^{\ominus}\ PyH^{\oplus}$$

3.1. Introduction

In 1969, Coates and Corrigan published in a brief paper[1] the use of pyridinium dichromate (PDC) as an alternative to the complex $CrO_3 \cdot 2Py$ in the oxidation of alcohols to aldehydes and ketones. Ten years later, Corey and Schmidt wrote a much-cited paper[2] in which full details on the oxidation of alcohols with PDC were provided. PDC is a stable bright-orange solid that is easily prepared by collecting the precipitate formed when pyridine is added to CrO_3 dissolved in a minimum amount of water. Normally, PDC transforms primary alcohols into aldehydes in a very effective manner.[3] On the other hand, when PDC is employed *in DMF as solvent*, it is able to transform *saturated* primary alcohols— but not allylic and benzylic ones—into the corresponding carboxylic acids in a very effective manner under essentially neutral conditions. The reaction is normally performed by adding PDC—either as a solid or dissolved in DMF—on a solution of the alcohol in DMF, and stirring the resultant mixture at room temperature.

> Very often, experimental descriptions of oxidations of alcohols to acids using PDC/DMF specify that *dry* DMF is used, and in some cases molecular sieves[4] are added, apparently for the purpose of avoiding epimerizations at the α position. On the other hand, many times the use of dry DMF is not specified in the experimental descriptions, and it is known that the oxidation can succeed employing a 40:1 DMF:water mixture.[5] Most oxidations of alcohols to acids proceed via an intermediate aldehyde hydrate that is

formed by reaction of an aldehyde with water. This raises the question of the potential intermediacy of an aldehyde hydrate, whose formation would need the presence of some amount of water, in oxidations of alcohols to acids using PDC/DMF.

Research needed

The mechanism of oxidation of primary alcohols to carboxylic acids with PDC in DMF must be studied. Particularly, the influence of water, molecular sieves, and some potential accelerants such as organic acids must be clarified, regarding reaction speed, yield, and possible epimerization at the α position.

Most times, oxidations of alcohols to acids with PDC/DMF are carried out at room temperature, although, on resistant substrates, the reaction mixture is sometimes heated at 40–50 °C.[6]

3.2. General Procedure for Oxidation of Aliphatic Primary Alcohols to Carboxylic Acids with Pyridinium Dichromate in Dimethylformamide. Method of Corey and Schmidt

From 1.2 to 10—typically 4–6—equivalents of pyridinium dichromate (MW = 376.2), either solid or dissolved as a ca. 0.2–1.3 M solution in DMF,[a] are added[b] to a ca. 0.04–0.3 M solution of 1 equivalent of the alcohol in DMF.[c] The resulting mixture is stirred at room temperature[d] until most of the starting alcohol is consumed.[e] Water is added[f] and the resulting mixture is extracted with an organic solvent such as EtOAc, Et$_2$O, or CH$_2$Cl$_2$. Optionally, the organic phase can be washed with brine. The organic solution is dried with MgSO$_4$ or Na$_2$SO$_4$ and concentrated, giving a crude carboxylic acid that may need further purification.

[a] The solubility of PDC in DMF is ca 0.9 g/mL at 25 °C.[2]

[b] Some heat is evolved during the mixing, therefore, particularly on a multigram scale, it may be advisable to cool the reaction mixture with an ice-water bath during the addition of PDC.

[c] *Dry* DMF is very often used and 3-Å molecular sieves[4] are sometimes added. On the other hand, the use of dry DMF is not specified in many experimental descriptions of this oxidation, and it is known that this oxidation can succeed employing a 40:1 DMF: water mixture.[5]

[d] For the oxidation of resistant substrates, it may be advisable to heat at 40–50 °C.[6]

[e] It normally takes between 3 hours and 2 days.

[f] In order to facilitate the extraction of the carboxylic acid from the aqueous phase, aqueous acetic acid or aqueous saturated NH$_4$Cl is sometimes added. In some cases, when the acid possesses a high solubility in water, it may be necessary to saturate the aqueous phase with sodium chloride.[7]

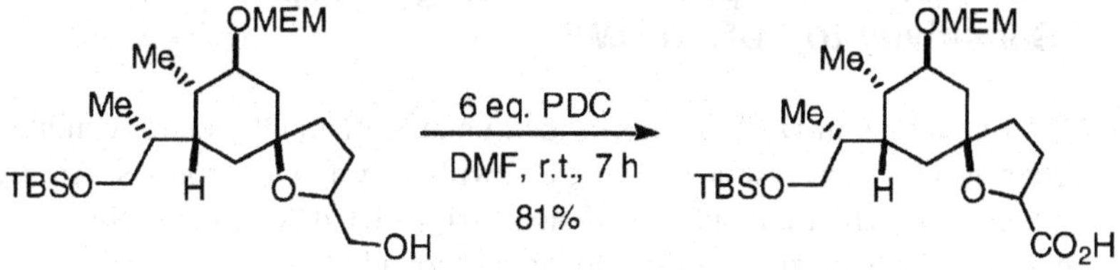

This difficult transformation was tried with many oxidants including Pt/O$_2$, Swern followed by Ag$_2$O, RuCl$_3$/cat. NaIO$_4$ and RuO$_4$. Best results were obtained employing PDC in DMF. [8]

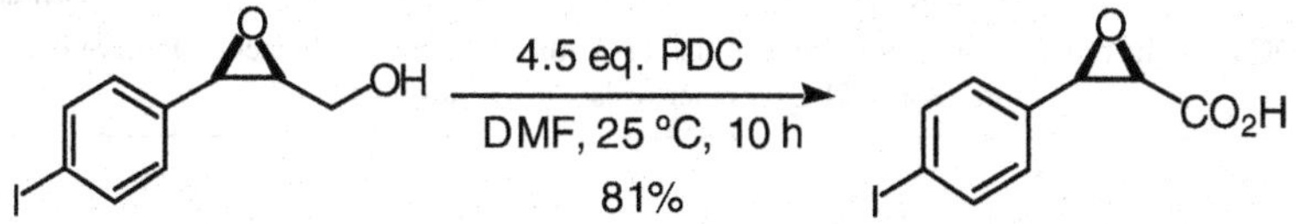

The use of RuCl$_3$/H$_5$IO$_6$ or PDC-Celite®, rather than PDC/DMF, provided the desired carboxylic acid, but in lower yields. [9]

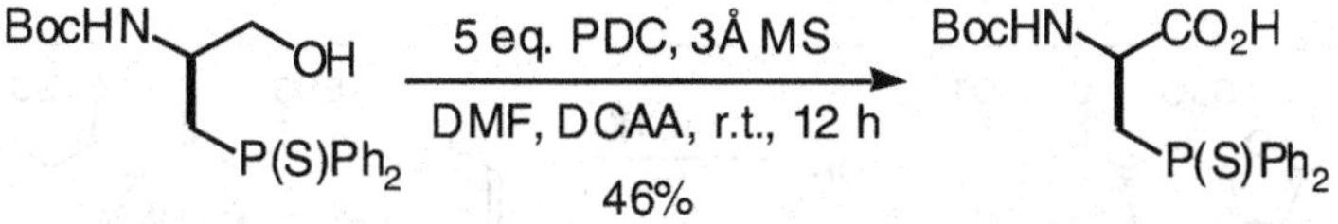

According to the authors, "Oxidation of the alcohol to the corresponding acid without racemization of the adjacent chiral center was a non-trivial challenge. The Moffatt oxidation has been used in similar situations where stereochemical integrity is an issue; this step, followed by further oxidation of the aldehyde produced using buffered permanganate, gave the corresponding Boc-protected amino acid in 94% ee. The best method identified for oxidation of the alcohol, however, was a pyridinium dichromate oxidation in the presence of molecular sieves, giving the product in > 99% ee." [4b]

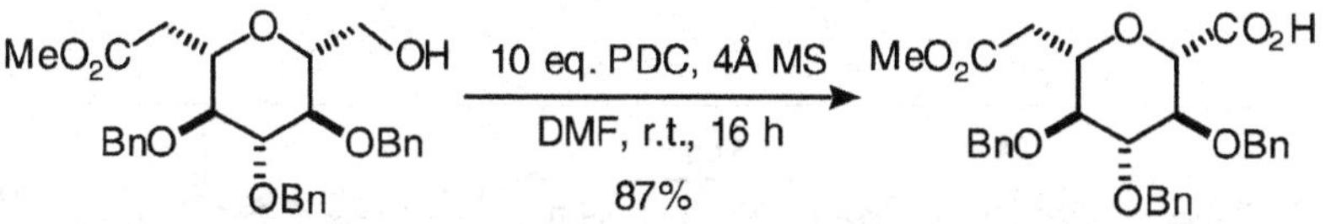

While a TEMPO-based oxidation fails to deliver a useful yield of carboxylic acid, the reaction succeeds employing PDC in DMF in the presence of molecular sieves. [10]

3.3. Functional Group and Protecting Group Sensitivity to PDC in DMF

As PDC oxidations in DMF are performed under essentially neutral conditions, protecting groups resist the action of PDC with very few exceptions. Although silyl ethers most often endure[11] oxidations with PDC/DMF, there are cases in which they are transformed in carbonylic compounds.[12]

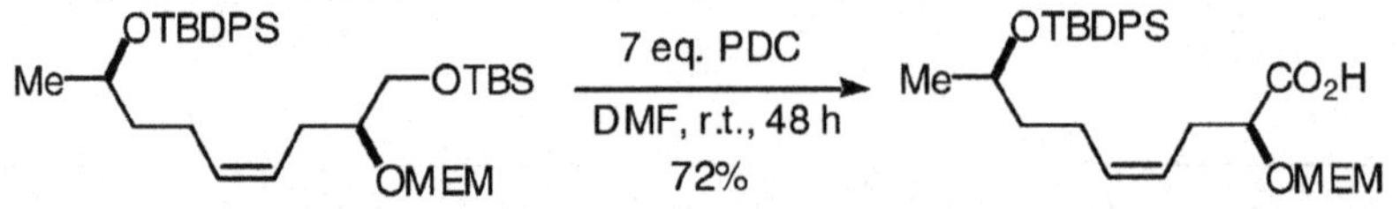

PDC in DMF produces the conversion of the TBS ether in a carboxylic acid. Observe that the more robust TBDPS ether resists the action of PDC.[12b]

Nevertheless, as the cleavage of silyl ethers with PDC in DMF is relatively slow, it is possible to oxidize free primary alcohols in the presence of unreacting silyl ethers.

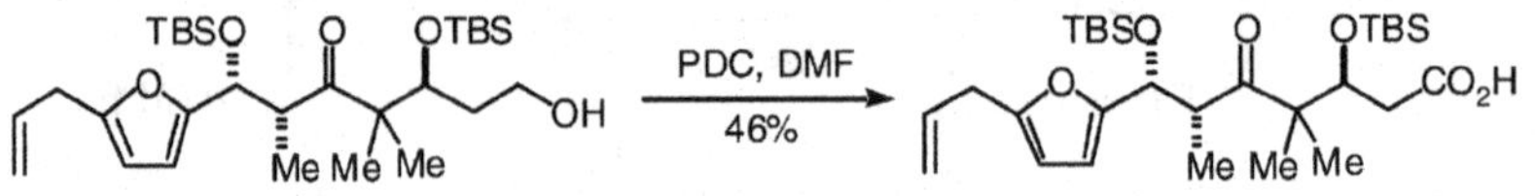

A primary alcohol is oxidized with PDC/DMF with no interference from two silyl ethers.[11c]

PDC is able to oxidize aldehydes to carboxylic acids,[13] secondary alcohols to ketones,[3] and lactols to lactones[14] under very mild conditions. Therefore, it is normally not possible to make selective oxidations of primary alcohols in the presence of those functional groups.

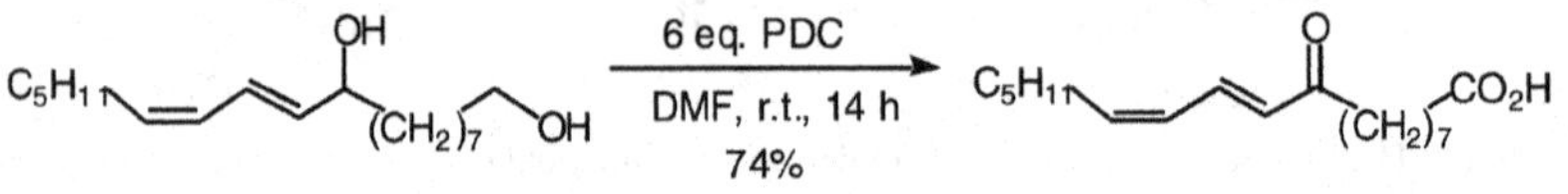

PDC in DMF is used for the simultaneous oxidation of a primary alcohol to carboxylic acid and a secondary alcohol to ketone.[15]

During the oxidation of certain diols with PDC in DMF, partial formation of lactones sometimes occurs.[16] This happens when the formation of the lactone is greatly favored for thermodynamical reasons.

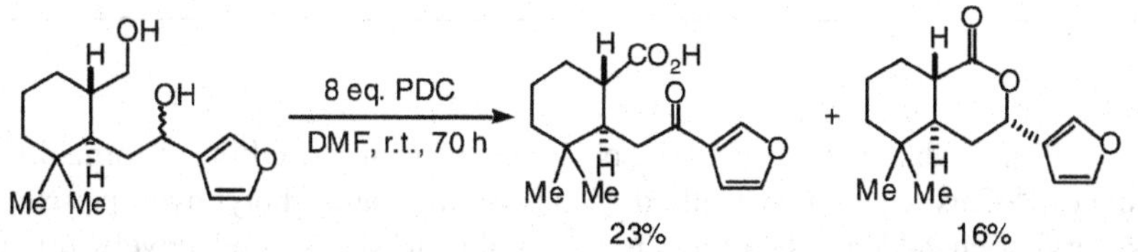

A diastereomeric mixture of two 1,5-diols is treated with PDC in DMF. The reaction products can be explained by the action of very subtle thermodynamical factors, whereby one of the diastereomers is uneventfully oxidized to a ketoacid while the other diastereomer is transformed into a lactone.[16b]

PDC is able to transform tertiary allylic alcohols into transposed enones,[17] a reaction that is normally carried out with PCC because it is more efficient.[18] Normally, it is possible to oxidize a primary alcohol in the presence of a tertiary allylic alcohol, because the reaction on the latter is slower.[19]

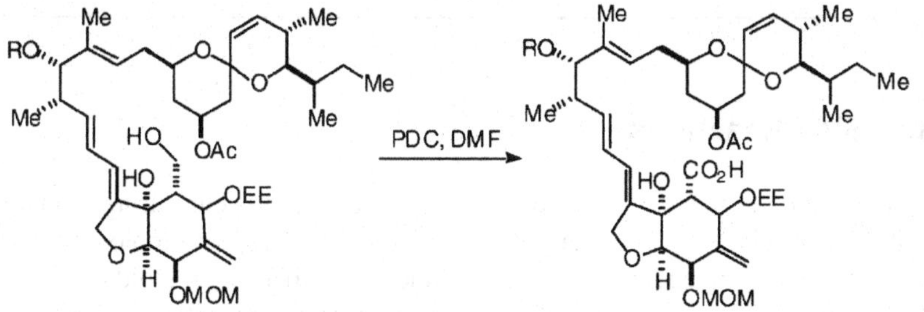

R= TBDPS-protected disacharide

An uneventful oxidation of a primary alcohol with PDC in DMF occurs, regardless of the presence of a tertiary allylic alcohol that could suffer an oxidative transposition to enone under more drastic conditions.[19]

Although PDC reacts with amines[20] and sulfides,[21] it is normally possible to oxidize selectively primary alcohols in the presence of tertiary[22] amines—which are less reactive than primary ones—or in the presence of sulfides.[23]

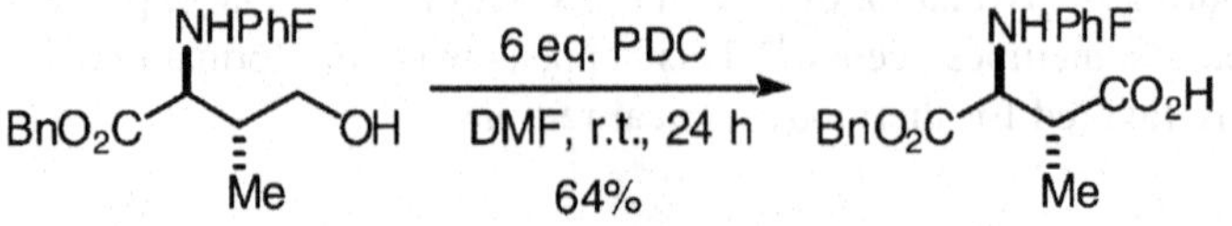

A primary alcohol is selectively oxidized with PDC in DMF in the presence of a secondary amine.[20a]

PDC is able to oxidize allylic positions in alkenes resulting in formation of enones.[24] This reaction normally demands heating; and t-butyl hydroperoxide[25] is very often added. Therefore, it is possible to oxidize selectively primary alcohols in the presence of alkenes,[26] because the oxidation of the former proceeds under milder conditions.

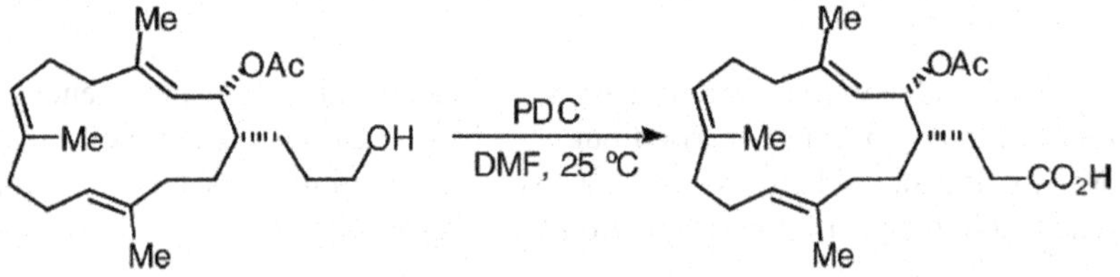

An uneventful oxidation of a primary alcohol to a carboxylic acid with PDC/DMF occurs, regardless of the presence of alkenes that could suffer oxidation to enones under more drastic conditions.[26a]

3.4. Side Reactions

Epimerization at the α position of carboxylic acids can occur during oxidations with PDC in DMF.[4b, 27] In fact, this side reaction can happen employing almost any other oxidizing agent, and it is sometimes very difficult to avoid completely. Sometimes, PDC in DMF is found to be particularly advantageous because it tends to produce less epimerization than other oxidants.[4b, 27b, 28]

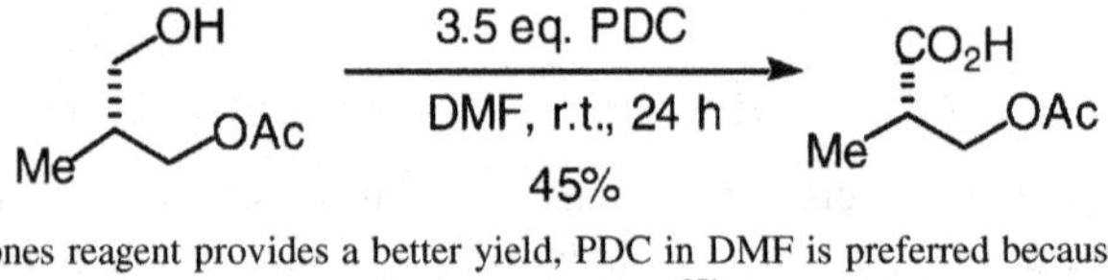

Although Jones reagent provides a better yield, PDC in DMF is preferred because it produces less racemization.[27b]

Normally, aliphatic primary alcohols are easily oxidized to carboxylic acids with PDC in DMF, while the oxidation of allylic and benzylic alcohols pauses at the aldehyde stage.[2] On the other hand, during the oxidation of aliphatic primary alcohols, carboxylic acids are occasionally obtained contaminated with minor amounts of aldehydes,[4b, 29] and in some rare cases, the oxidation of aliphatic alcohols stops at the aldehyde stage with no carboxylic acid being isolated.[30]

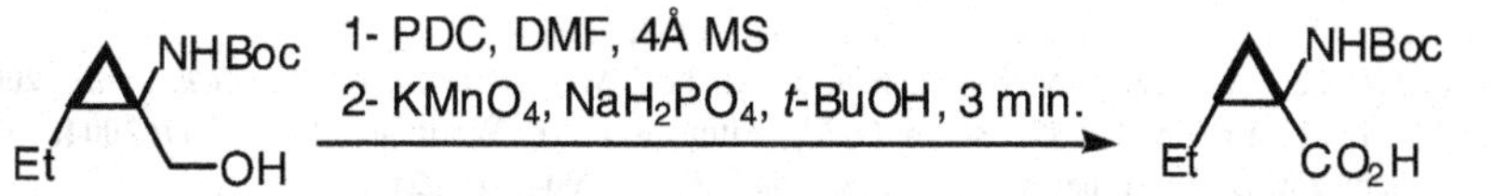

This is an uncommon case in which a primary aliphatic alcohol is treated with PDC in DMF resulting in an unfinished oxidation to carboxylic acid that must be completed by treatment with potassium permanganate. The electronic similitude between alkenes and cyclopropanes may explain this fact, because the primary alcohol in this molecule possesses a certain allyl-like character and it is known that allyl alcohols are oxidized to aldehydes by PDC in DMF.[30b]

Occasionally, dimeric esters are formed during oxidation of primary aliphatic alcohols with PDC in DMF.[7] This can be mitigated by the use of excess of PDC and avoiding high temperatures.[7]

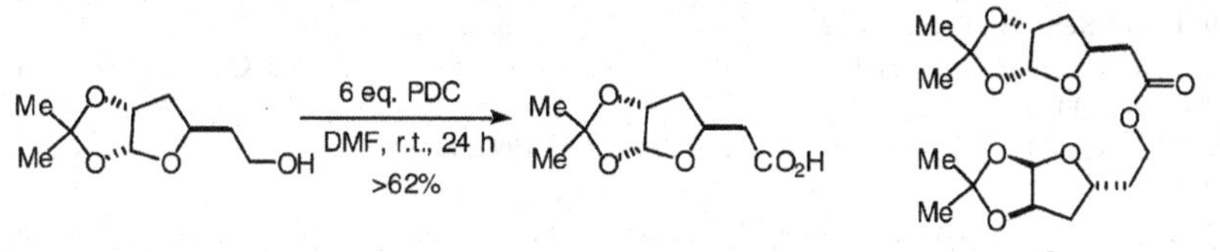

A good yield of carboxylic acid is obtained by performing the oxidation at 18 °C and using 6 equivalents of DMF. A 10% yield of the undesired dimeric ester **8** is formed if the reaction is run at higher temperature and employing less equivalents of DMF.[7]

3.5. References

1 Coates, W. M.; Corrigan, J. R.; *Chem. Ind.* **1969,** 1594.

2 Corey, E. J.; Schmidt, G.; *Tetrahedron Lett.* **1979,** *20,* 399.

3 Tojo, G.; Fernández, M. in *Basic Reactions in Organic Synthesis. Oxidation of Alcohols to Aldehydes and Ketones: A Guide to Current Common Practice*, G. Tojo, Ed.; Springer: New York, 2006, p. 28.

4 a) Kinoshita, M.; Arai, M.; Ohsawa, N.; Nakata, M.; *Tetrahedron Lett.* **1986**, *27*, 1815.
   b) Porte, A. M.; van der Donk, W. A.; Burgess, K.; *J.Org.Chem.* **1998**, *63*, 5262. c) Sato, T.;
   Aoyagi, S.; Kibayashi, C.; *Org.Lett.* **2003**, *5*, 3839.

5 Chen, B.; Ko, R. Y. Y.; Yuen, M. S. M.; Cheng, K.-F.; Chiu, P.; *J.Org.Chem.* **2003**, *68*, 4195.

6 a) Kurokawa, N.; Ohfune, Y.; *J.Am.Chem.Soc.* **1986**, *108*, 6041. b) Manesis, N. J.; Goodman, M.;
   J.Org.Chem. **1987**, *52*, 5342. c) Azuma, H.; Takao, R.; Niiro, H.; Shikata, K.; Tamagaki, S.;
   Tachibana, T.; Ogino, K.; *J.Org.Chem.* **2003**, *68*, 2790.

7 Gillard, F.; Heissler, D.; Riehl, J.-J.; *J.Chem.Soc. Perkin Trans. I* **1988**, 2291.

8 Ireland, R. E.; Maienfisch, P.; *J.Org.Chem.* **1988**, *53*, 640.

9 Boger, D. L.; Patane, M. A.; Zhou, J.; *J.Am.Chem.Soc.* **1994**, *116*, 8544.

10 van Well, R. M.; Overkleeft, H. S.; van Boom, J. H.; Coop, A.; Wang, J. B.; Wang, H.; van der
   Marel, G. A.; Overhand, M.; *Eur.J.Org.Chem.* **2003**, 1704.

11 a) Fürstner, A.; Albert, M.; Mlynarski, J.; Matheu, M.; DeClercq, E.; *J.Am.Chem.Soc.* **2003**,
   125, 13132. b) Schinzer, D.; Böhm, O. M.; Altmann, K.-H.; Wartmann, M.; *Synlett* **2004**, 1375.
   c) Schinzer, D.; Bourguet, E.; Ducki, S.; *Chem.Eur.J.* **2004**, *10*, 3217.

12 a) Denmark, S. E.; Hammer, R. P.; Weber, E. J.; Habermas, K. L.; *J.Org.Chem.* **1987**, *51*, 165.
   b) Solladié, G.; Almario, A.; *Tetrahedron: Asymmetry* **1995**, *6*, 559. c) Solladié, G.; Urbano, A.;
   Stone, G. B.; *Tetrahedron Lett.* **1993**, *34*, 6489.

13 a) Cachet, X.; Deguin, B.; Tillequin, F.; Rolland, Y.; Koch, M.; *Helv.Chim.Acta* **2000**, *83*, 2812.
   b) Sato, K.; Yoshimura, T.; Shindo, M.; Shishido, K.; *J.Org.Chem.* **2001**, *66*, 309. c) Lesuisse, D.;
   Gourvest, J.-F.; Albert, E.; Doucet, B.; Hartmann, C.; Lefrançois, J.-M.; Tessier, S.; Tric, B.;
   Teutsch, G.; *Bioorg.Med.Chem.Lett.* **2001**, *11*, 1713. d) Jain, R. P.; Williams, R. M.; *J.Org.Chem.*
   2002, *67*, 6361.

14 Deng, Y.; Salomon, R. G.; *J.Org.Chem.* **2000**, *65*, 6660.

15 Tsuboi, S.; Maeda, S.; Takeda, A.; *Bull.Chem.Soc.Jpn* **1986**, *59*, 2050.

16 a) Weinges, K.; Lernhardt, U.; *Lieb.Ann.Chem.* **1990**, 751. b) Ihara, M.; Suzuki, S.; Taniguchi, N.;
   Fukumoto, K.; *J.Chem.Soc. Perkin Trans. I* **1993**, 2251.

17 a) Nagaoka, H.; Baba, A.; Yamada, Y.; *Tetrahedron Lett.* **1991**, *32*, 6741. b) Liotta, D.; Brown, D.;
   Hoekstra, W.; Monahan III, R.; *Tetrahedron Lett.* **1987**, *28*, 1069.

18 a) Dauben, W. G.; Michno, D. M.; *J.Org.Chem.* **1977**, *42*, 682. b) Bacigaluppo, J. A.; Colombo,
   M. I.; Zinczuk, J.; Huber, S. N.; Mischne, M. P.; Rúveda, E. A.; *Synth.Commun.* **1991**, *21*, 1361.

19 Fraser-Reid, B.; Barchi Jr., J.; Faghih, R.; *J.Org.Chem.* **1988**, *53*, 923.

20 Zhu, X.; Greig, N. H.; Holloway, H. W.; Whittaker, N. F.; Brossi, A.; Yu, Q.-sheng; *Tetrahedron
   Lett.* **2000**, *41*, 4861.

21 Mangalam, G.; Sundaram, S. M.; *J.Ind.Chem.Soc.* **1991**, *68*, 77.

22 Kazmierski, W. M.; *Int.J.Pept.Protein.Res.* **1995**, *45*, 241.

23 a) Jones, D. N.; Meanwell, N. A.; Mirza, S. M.; *J.Chem.Soc. Perkin Trans. I* **1985**, 145.
   b) Balsamo, A.; Benvenuti, M.; Giorgi, I.; Macchia, B.; Macchia, F.; Nencetti, S.; Orlandini, E.;
   Rosello, A.; Broccali, G.; *Eur.J.Med.Chem.* **1989**, *24*, 573. c) Urbanski, M. J.; Chen, R. H.;
   Demarest, K. T.; Gunnet, J.; Look, R.; Ericson, E.; Murray, W. V.; Rybczynski, P. J.; Zhang, X.;
   Bioorg.Med.Chem.Lett. **2003**, *13*, 4031.

24 a) Schultz, A. G.; Lavieri, F. P.; Macielag, M.; Plummer, M.; *J.Am.Chem.Soc.* **1987**, *109*, 3991.
   b) Schultz, A. G.; Plummer, M.; Taveras, A. G.; Kullnig, R. K.; *J.Am.Chem.Soc.* **1988**, *110*, 5547.

25 Chidambaram, N.; Chandrasekaran, S.; *J.Org.Chem.* **1987**, *52*, 5048.

26 a) Marshall, J. A.; Gung, W. Y.; *Tetrahedron Lett.* **1988**, *29*, 3899. b) Kikuchi, H.; Sasaki, K.;
   Sekiya, J.'ichi; Maeda, Y.; Amagai, A.; Kubohara, Y.; Oshima, Y.; *Bioorg.Med.Chem.* **2004**, *12*,
   3203. c) Hu, T.; Panek, J. S.; *J.Am.Chem.Soc.* **2002**, *124*, 11368.

27 a) Martin, S. F.; Dappen, M. S.; Dupré, B.; Murphy, C. J.; Colapret, J. A.; *J.Org.Chem.*
   1989, *54*, 2209. b) Allali, H.; Tabti, B.; Alexandre, C.; Huet, F.; *Tetrahedron: Asymmetry*
   2004, *15*, 1331.

28 Nakamura, K.; Baker, T. J.; Goodman, M.; *Org.Lett.* **2000**, *2*, 2967.

29 Basak, A.; Nag, A.; Bhattacharya, G.; Mandal, S.; Nag, S.; *Tetrahedron: Asymmetry* **2000,** *11,* 2403.

30 a) Avery, M. A.; Fan, P.; Karle, J. M.; Bonk, J. D.; Miller, R.; Goins, D. K.; *J.Med.Chem.* **1996,** *39,* 1885. b) Charette, A. B.; Côté, B.; *J.Am.Chem.Soc.* **1995,** *117,* 12721.

Heyns Oxidation

O₂, Pt

4.1. Introduction

In the 1940s, Heyns *et al.* reported[1] that 2-keto-L-gulonic acid (**10**) can be obtained in high yield by bubbling oxygen through an aqueous solution of L-sorbose (**9**) containing suspended finely divided platinum.

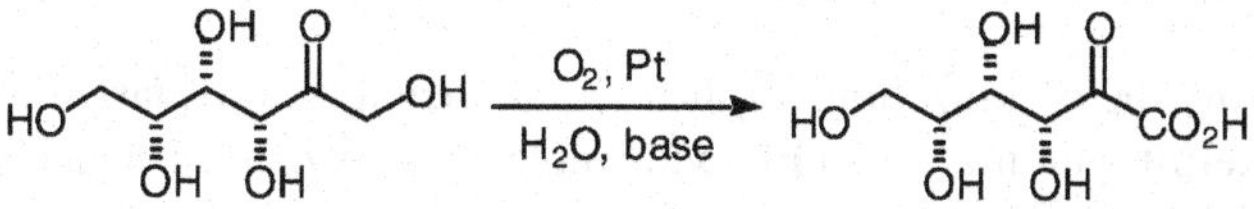

L-sorbose (**9**)   2-keto-L-gulonic acid (**10**)

This remarkable transformation involves the selective oxidation of one primary alcohol in a molecule containing two primary alcohols and three secondary ones. Because of its mildness and selectivity, these reaction conditions were tested in other sugars and resulted in highly efficient oxidations of primary alcohols to acids.[2]

Although Heyns must be credited for his early very important contributions to the development of the oxidation of alcohols using oxygen in the presence of catalytic platinum, it must be mentioned that the catalytic effect of platinum in the oxidation of cinnamic alcohol with oxygen was already noted by Strecker in 1855.[3] Other early contributions to the oxidation of alcohols with oxygen in the presence of platinum were made by von Gorup-Besanez,[4] Dafert,[5] and Grimeaux.[6]

The experimental data,[7] for the platinum-catalyzed oxidation of primary alcohols to carboxylic acids using molecular oxygen, are compatible with a mechanism, shown below, beginning with complexation of the alcohol with an active site of the catalyst and transfer of hydrogen atoms from the alcohol to the surface of the platinum particles. This step—which is the reversal of catalytic hydrogenation of aldehydes—leads to an intermediate aldehyde and hydrogen being adsorbed on platinum particles. Next, the hydrogen atoms react with oxygen under platinum catalysis resulting in formation of water, probably via formation of hydrogen peroxide.[7g] The intermediate aldehyde is hydrated resulting in formation of a *gem*-diol that is further oxidized to carboxylic acid through a pathway resembling the transformation of alcohol into aldehyde.

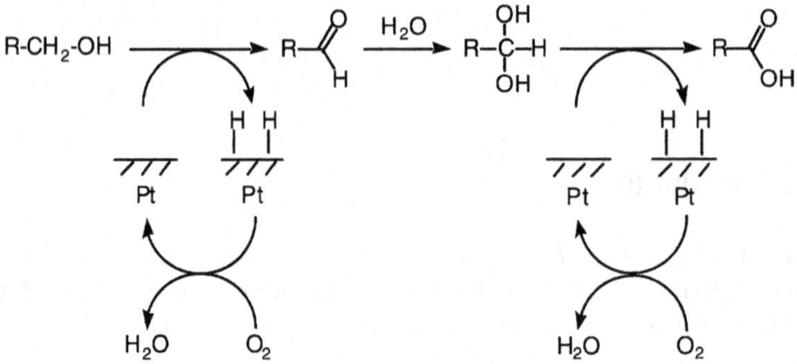

During the optimization[1] of the platinum-catalyzed oxidation of L-sorbose (**9**) to 2-keto-L-gulonic acid (**10**) with molecular oxygen in aqueous solution, Heyns *et al.* established the following experimental facts:

- The pH must be neutral or slightly basic. The oxidation is very slow under acidic conditions. Although basic pH results in acceleration of the reaction, the pH must be kept below 11 to avoid base-catalyzed decomposition of sugars. The reaction runs satisfactorily between pH 8 and 10. As the reaction proceeds, there is a decrease of pH due to generation of carboxylic acid, until a value of ca. 2.4 is reached. This can lead to cessation of the reaction unless a base is added. In order to keep the reaction advancing, the pH can be adjusted to a value of 8–10 by the periodic addition of base. Alternatively, 1 to 3 equivalents of base can be added at the outset resulting in a high pH that decreases as the reaction proceeds and produces a deceleration of reaction speed that helps to prevent overoxidations. Dibasic sodium phosphate, NaOAc, potassium oxalate, Na_2CO_3, $NaHCO_3$, and KOH are suitable bases.
- In order to reach an appropriate reaction velocity, the oxygen must be thoroughly mixed into the medium by bubbling it through a solution subject to energic shaking or stirring. Air can be employed instead of pure oxygen at the cost of a much decreased oxidation rate, although the milder oxidation conditions can lead eventually to better yield of acid.
- Not surprisingly, heating causes an increase of velocity, 60–70 °C being a suitable reaction temperature, although better yield can be obtained at 28–35 °C because of milder reaction conditions.

- The concentration of L-sorbose (**9**) must not exceed 5–6%.
- Metallic platinum must be present in finely divided form. Black platinum generated by reduction of PtO$_2$ with hydrogen can be employed. Alternatively, platinum deposited on a carbon support can also be used; 5–10% platinum on active carbon provides best yields.
- For best results, a generous amount of catalyst must be employed. For example, in order to oxidize 18 g of L-sorbose (**9**), good yields are obtained using 10 g of 10% platinum on carbon. Fortunately, the costly catalyst can be recovered and reused[1b, 8] more than 20 times with no noticeable loss of activity.

Heyns *et al.* tried many catalysts for the oxidation of L-sorbose (**9**) with oxygen, including around 30 different elements. Apart from platinum, only palladium and osmium delivered some 2-keto-L-gulonic acid (**10**).[1b]

After the initial studies by Heyns *et al.* on the oxidation of L-sorbose (**9**), the platinum-catalyzed oxidation of primary alcohols with oxygen was tested on numerous substrates by many researchers. As a rule, subsequent studies confirmed the early observations; nonetheless, small changes of the original protocol very often allow some improvements in the yields.

Solvent

These reactions are most often carried out in water, in which case the alcohol must have a certain solubility for the reaction to proceed. Ideally, the alcohol must be completely dissolved in water, although the reaction very often succeeds employing a suspension of the alcohol.[9] In any case, the reaction medium must be completely uniform and the catalyst must be able to move without obstruction inside the liquid. Traces of oil droplets in water can cause the coagulation of catalyst particles and arrest the continuation of the oxidation.[2]

When an alcohol possessing a low melting point is oxidized as a suspension in water, the reaction temperature must not exceed the melting point in order to avoid clotting of the catalyst.[9c]

During this oxidation, oxygen is bubbled through a suspension of starting alcohol in water. For greater reaction velocity, it is better to use high temperature, but it must not exceed the melting point of the starting alcohol, because the resulting oily drops would cause coagulation of the catalyst particles and interruption of the oxidation.[9c]

Heyns oxidation can be performed in an organic solvent, such as EtOAc,[10] glacial acetic acid,[11] or heptane,[13a] instead of water. In such cases an aldehyde is obtained, something that is not surprising because, in the absence of water, the aldehyde cannot be transformed into a *gem*-diol that could be further oxidized to acid. On the other hand, in regular Heyns oxidations carried out in water, the addition of an organic cosolvent can sometimes be beneficial because it can promote better mixing of reagents. Solvent combinations employed for this purpose include: acetone and water;[12] dioxane and water;[13] EtOAc and water;[14] diglyme and water;[15] isopropyl alcohol, EtOAc, and water;[16] and isopropyl alcohol, acetone, and water.[17] The addition of a surfactant such as Dow Corning Antifoam A[18] or sodium lauryl sulfate[15] may help to facilitate the solubilization of the alcohol and increase the yield.

Supercritical carbon dioxide can be used as solvent in Heyns oxidations.[19]

pH

Figure 2 shows the correlation between pH and reaction speed during the oxidation of L-sorbose (**9**) to 2-keto-L-gulonic acid (**10**).[20] While there is hardly any oxidation below pH 6, there is an approximate sixfold increase in velocity from pH 7 to pH 9. On the other hand, there is a decrease in selectivity of oxidation of the primary alcohol with increasing pH, due to base-catalyzed

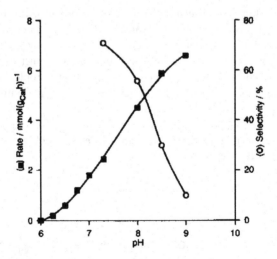

Figure 2. Influence of pH on average reaction rate and selectivity in the oxidation of L-sorbose (**9**) to 2-keto-L-gulonic acid (**10**) measured at 30% conversion (5 mass% Pt/Al_2O_3, water, 50°C, O_2). Selectivity refers to the percentage of consumed L-sorbose (**9**) transformed into 2-keto-L-gulonic acid (**10**). Reproduced by permission of The Royal Society.

decompositions. Because of these factors, oxidation of sugars is normally better performed at pH ca. 7–8.

Heyns oxidations are normally carried out in the presence of $NaHCO_3$, to avoid acidification due to generation of a carboxylic acid, which would cause a very substantial decrease in oxidation speed. Sometimes, all the base is added at once at the beginning of the oxidation, although it may be more satisfactory to fine-tune the pH continuously by the sequential addition of several portions of base. Other bases less commonly employed include NaOAc,[1a, 21] KOH,[1a] potassium oxalate,[1a] potassium oxalate plus $NaHCO_3$,[1a] phosphate buffer,[1a, 22] $NaHCO_3$ plus K_2CO_3 (pH 7.5–9.0),[9a] $KHCO_3$,[23] $BaCO_3$,[7h] K_2CO_3,[24] and $NaHCO_3$ plus Na_2CO_3 (pH 8.8).[25]

Secondary alcohols can also be oxidized under Heyns conditions,[26] although generally at a much slower rate than primary ones. On the other hand, at acidic or neutral pH, in which primary alcohols are slowly oxidized and their oxidation tends to stop at the aldehyde stage,[2] the oxidation of secondary alcohols may occur at a substantial speed. Therefore, in order to accomplish good selectivity in the oxidation of primary alcohols to acids in the presence of secondary alcohols, it is advisable to operate at a basic pH. On the other hand, Heyns oxidation of primary alcohols to carboxylic acids can occasionally be successful at low pH.[25, 27]

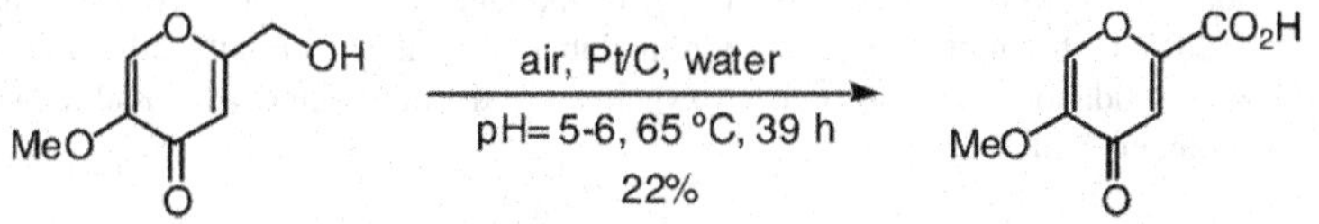

The obtention of this carboxylic acid under Heyns conditions must be done under acidic pH—something that is unusual in Heyns oxidations—due to sensitivity of the pyranone ring to base. The acidic conditions lead to a decrease in oxidation rate, which results in moderate yield of acid in spite of prolonged reaction time in the presence of generous quantities of catalyst (1.25 g Pt/C per g starting alcohol).[27a]

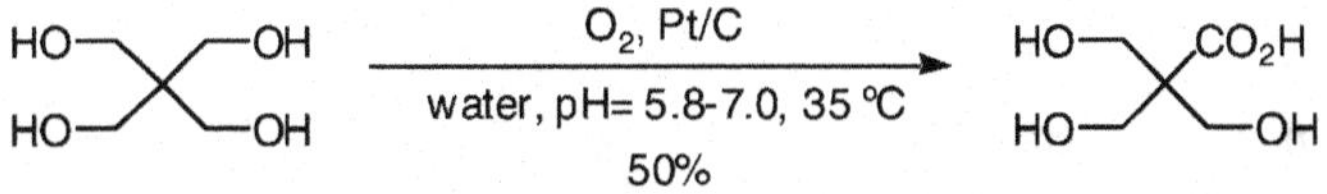

In this remarkable monooxidation of a symmetrical compound possessing four primary alcohols, it is important to run the reaction under mild acidic conditions and moderate temperature to avoid overoxidation.[27b]

It must be mentioned that successful Heyns oxidations of primary alcohols are sometimes carried out with no addition of base.[12c, 16, 28]

Oxygen

The oxidation of alcohols with oxygen and platinum is normally quite slow. Therefore, it is usually practical to speed up the reaction by (a) using high temperature—normally close to the boiling point of water; (b) bubbling[8] pure gaseous oxygen, rather than air, in the most energetic way; (c) and stirring[8, 29] or shaking the reaction medium in the most vigorous manner.

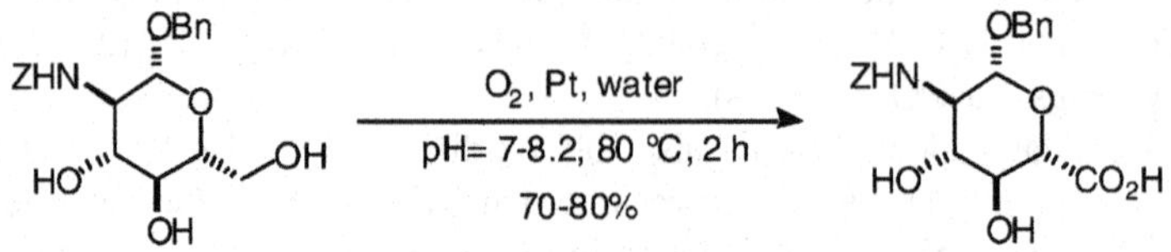

In this oxidation it is possible to raise the yield from 40% to 70–80%, simply by quickening the stirring speed and the bubbling of oxygen.[8]

It can be proved that above a certain stirring velocity no further acceleration of oxidation occurs.[71] Likewise, increasing the amount of oxygen can in fact produce a reduction of oxidation rate due to oxygen blocking the active sites of the catalyst,[30] something that is unfortunate because it prevents an efficient acceleration of this oxidation by working under high oxygen pressure.[2, 31] As these severe conditions are difficult to implement in a normal laboratory, it is normally advisable to run Heyns oxidations with as much oxygen and stirring/shaking as possible with the equipment available.

Research needed

Oxygen is not very efficient in removing hydrogen atoms from the surface of platinum particles. Alternative oxidants such as methylene blue[7b] or hydrogen peroxide[9a] were briefly tested. A more efficient and practical secondary oxidant than gaseous oxygen is needed.

Platinum

Heyns oxidations are performed using either platinum deposited on a carbon support or black platinum generated by reduction of Adams' catalyst, a highly reactive form of hydrated platinum dioxide. It must be mentioned that the effectivity of a certain platinum on carbon catalyst greatly varies depending on the kind of carbon support employed.[1b, 32] Some authors claim that reduced Adams' catalyst is superior to platinum on carbon.[7j, 26c, 33] Although reduced Adams' catalyst is expected to provide more consistent results, certain kinds of platinum on carbon may work more efficiently, especially on a platinum weight basis.

> Platinum deposited on Al_2O_3[7m, 34] and platinum-containing hyper-cross-linked polystyrene[35] have also been used.

WARNING: during Heyns oxidations in which black platinum is prepared *in situ* by reduction of platinum dioxide with H_2, the gaseous hydrogen must be completely evacuated and replaced by an inert gas like nitrogen or argon before oxygen is introduced. The simultaneous presence of hydrogen and oxygen in the reaction flask must be rigorously avoided, particularly considering that platinum is present. Platinum may catalyze an explosive reaction of hydrogen with oxygen.

Poisons

In common with other platinum-catalyzed reactions, such as catalytic hydrogenation, the platinum catalyst may be poisoned by traces of certain compounds that easily block the catalytic sites. As expected, sulfur compounds such as dithioacetals[36] and H_2S[7j] strongly inhibit Heyns oxidations. Some metallic ions such as calcium and aluminum cations also inhibit the oxidation. The inhibition produced by Ca^{2+} is particularly strong and common.[2] Stagnant distilled water normally contains Ca^{2+} ions dissolved off the glass cylinders, and therefore it is recommended that Heyns oxidation be done in *freshly* bidistilled water. By employing freshly bidistilled water instead of stagnant distilled water, it was possible to shorten the oxidation of isopropyl alcohol to acetone from 7000 to 50 minutes.[13a]

Amines poison the platinum catalyst because they strongly bind to the catalytic sites.[7j, 27c, 37]

> Interestingly, amines in very small amounts can favor Heyns oxidations because of the accelerating effect provided by a local increase of the pH on the surface of platinum particles. The quantity of amine must be finely adjusted in order to keep this beneficial effect while sufficient active sites remain available for complexation of substrate.[7m]
> The inhibitory effect of phenols is open to debate.[9a,b, 25, 27]

When the oxidation is performed using compressed air, poisoning from oil coming from pumps is possible. Therefore, it is recommended that air be washed,[37] for example, by passing through concentrated sulfuric acid.[7j] This problem is not so common when compressed oxygen is employed.

Hydrogen adsorbed on the platinum particles inhibits the oxidation.[9b] That is why, when clean platinum is used, the velocity of the oxidation is very high on the first contact with the alcohol. This is followed by a slower oxidation rate while the oxygen has to remove the hydrogen from platinum particles.[7l] In fact, excess of platinum in the absence of oxygen would have been a magnificent oxidant were it not for its price.

When black platinum is generated by treatment of PtO_2 with hydrogen, the resulting particles are covered with hydrogen, and it is important to remove rigorously this hydrogen to achieve a high initial oxidation rate.

The products of Heyns oxidation are sometimes adsorbed on the platinum surface stronger than the substrate.[7k] In such cases of product-poisoning of the catalyst, it is very difficult to achieve full conversion of the starting compound, and, for best yields, it is better not to prolong the oxidation. An efficient complexation of product with active sites, which hinder the oxidation, can depend on very subtle structural features.[34] Therefore, the speed of the oxidation of a certain primary alcohol may be very difficult to anticipate, based on comparisons with similar substrates.

Selectivity

The most remarkable property of Heyns oxidation is its ability to selectively oxidize primary alcohols in the presence of secondary ones.[32a, 34, 38] This allows, for example, very effective transformations in unprotected polyols.

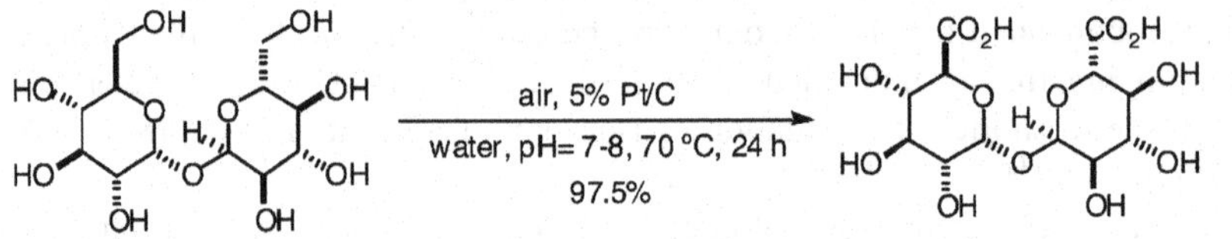

Heyns oxidation allows a selective and almost quantitative oxidation of two primary alcohols in a molecule containing six unprotected secondary alcohols.[39]

On the other hand, it must be mentioned that under Heyns' conditions secondary alcohols can also suffer oxidation.[2, 13a, 26a,c-f, 32a] In fact, during the oxidation of many sugars, secondary alcohols react resulting in oxidative breakages of carbon–carbon bonds that lead to low-molecular-weight compounds that remain undetected in many experiments. Very often only the product, resulting from the oxidation of the primary alcohol and possessing higher molecular weight, is detected, leading to the view that the reaction is highly selective. When care is taken to isolate low-molecular-weight products, a picture emerges in which, although the major pathway is the selective oxidation of the primary alcohol, substantial oxidation of secondary alcohols also occurs.[34]

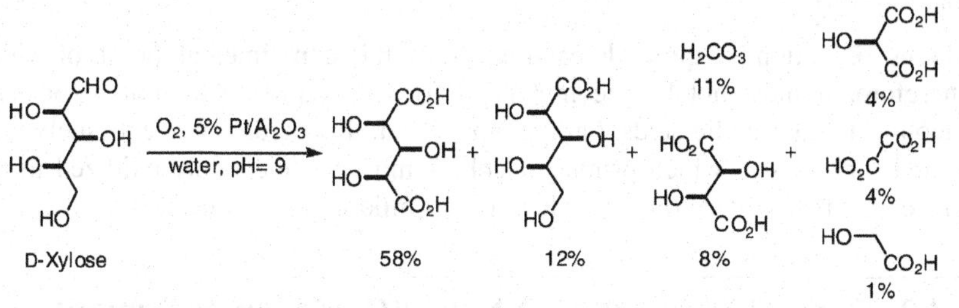

During an oxidation of D-xylose, optimized for obtention of the dicarboxylic acid resulting from reaction of both the aldehyde and the primary alcohol, many low-molecular-weight compounds, resulting from oxidative breakage of secondary alcohols, are obtained.[34]

As expected, higher selectivity on oxidation of primary alcohols in the presence of secondary ones is achieved operating under milder conditions.[28a] While the oxidation of primary alcohols is accelerated at basic pH and is greatly suppressed under acidic conditions, secondary alcohols are oxidized at a substantial rate under acidic conditions. Therefore, it is advisable to operate at basic pH to achieve an optimum selectivity in the oxidation of primary alcohols.

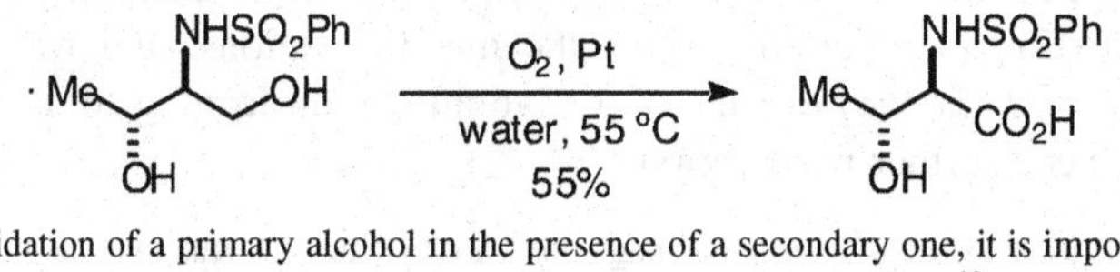

In this oxidation of a primary alcohol in the presence of a secondary one, it is important not to increase the temperature to maintain good selectivity.[28a]

Due to the mildness of Heyns oxidation, it is often possible to selectively oxidize certain primary alcohols in molecules possessing more than one alcohol of this kind.[40]

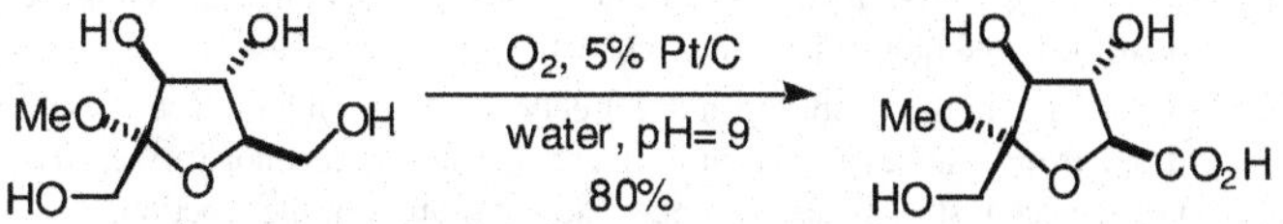

The less hindered primary alcohol is selectively oxidized in 80% yield in a molecule containing two primary and two secondary alcohols.[40b]

Application

Heyns oxidation is quite demanding from the experimental point of view; therefore, it must not be recommended for the routine oxidation of primary alcohols to carboxylic acids. On the other hand, this oxidation is extremely well suited for cases in which primary alcohols must be selectively oxidized in the presence of secondary ones or when a very mild oxidant is needed.

4.2. General Procedure for Heyns Oxidation of Primary Alcohols to Carboxylic Acids

Oxygen[a] is bubbled[b] over a vigorously[c] stirred or shaken suspension[d] of black[e,f] platinum (ca. 0.6–2.1 equivalents) or platinum on carbon[f,g] (ca. 95–210 g/mol alcohol) in hot[h] water[i], containing 1 equivalent of alcohol,[j] in a concentration of ca. 0.01–0.25 mol/liter, and approximately 1–11 equivalents[k] of $NaHCO_3$.[l] When most of the starting compound is consumed,[m] the catalyst is filtered using either paper or a pad of Celite®. The basic aqueous solution can be optionally washed with an organic solvent such as EtOAc. The aqueous phase is acidified with hydrochloric or sulfuric acid. In some cases this produces crystallization of the acid that can be filtered. Alternatively, the acid can be extracted with an organic solvent such as EtOAc, and the organic solution can be dried (Na_2SO_4 or $MgSO_4$) and concentrated, giving a crude acid that may need further purification. Optionally, the basic aqueous solution containing the acid can be treated with an acidic resin in order to remove the metallic cations, and water can be removed by (*in vacuo*) distillation or lyophilization leading to a crude acid that may need further purification.

[a] Air can be used instead of oxygen, although this leads to decreased reaction rate. It is advisable to wash the air by passing through concentrated sulfuric acid, to avoid poisoning of the catalyst from residues of oil from an air compressor or from amines. Poisoning is not so likely when oxygen from a cylinder is employed.

[b] Bubbling must be as energetic as possible to achieve optimum oxidation speed.

[c] The experimental setup must guarantee maximum mixing and contact among the solution, the catalyst particles, and oxygen. Oxygen can be bubbled through the reaction medium in a standard setup with magnetic or mechanical stirring. Alternatively, the oxygenation can be made using a hydrogenation apparatus. WARNING: in such case it is *extremely important* to remove any trace of hydrogen before oxygen is introduced, because platinum may catalyze an explosion of oxygen in contact with hydrogen. The mixing must be done so that most of the catalyst particles remain in suspension in the liquid rather than deposited on the walls of the flask.

[d] The catalyst particles must be able to move freely inside the liquid. The presence of even small oily drops from residues of organic solvents or starting alcohol very easily causes the coagulation of the catalyst particles, leading to near cessation of the reaction.

[e] Black platinum is normally prepared *in situ* by reduction of Adams' catalyst—a highly reactive form of hydrated platinum dioxide—with gaseous hydrogen. Hydrogen is admitted into a stirred or shaken supension of Adams' catalyst in water. This produces a change from

a heavy granular powder of brown Adams' catalyst to a much finer suspension of black particles of metallic platinum, the so-called black platinum. In the absence of organic matter or added salts, the generation of black platinum normally takes from several seconds to 2 or 3 minutes.[41] WARNING: after the preparation of black platinum, it is very important to replace the unreacted hydrogen by an inert gas like nitrogen or argon before bringing oxygen in, because platinum may induce an explosion if hydrogen is allowed to be in contact with oxygen. Black platinum *freshly* prepared by hydrogenation of PtO_2 is covered by hydrogen atoms that block the active sites. Failure to remove these hydrogen atoms may lead to an induction period in which no oxidation of alcohols takes place while oxygen cleans the catalytic particles.

[f] Both black platinum and platinum on carbon can be reused[1b, 8] in many oxidations with no substantial loss of activity. Due to the very easy poisoning of catalyst with traces of sulfur compounds, amines, and certain inorganic cations such as Ca^{2+}, all components of the reaction mixture must be of the highest purity.

[g] Normally, 5% or 10% platinum on carbon is used. There may be substantial differences in performance depending on carbon support.[1b, 32] Therefore, it is important to disclose the source and type of carbon-supported platinum when reporting a certain experiment.

[h] Water at a temperature between 50 and 90 °C is normally employed. The reaction can be performed at room temperature, resulting in better selectivity on the oxidation of primary alcohols at the cost of much reduced oxidation rate.

[i] Stale distilled water stored in glass containers is often contaminated with Ca^{2+} ions leached from the walls that can retard substantially the reaction by poisoning of the catalyst.[2] Therefore, it is recommended to employ *freshly* bidistilled water.

[j] Ideally, the alcohol must be completely dissolved in water. The oxidation may succeed with the alcohol suspended in water, provided that the alcohol possesses certain solubility. The segregation of some liquid alcohol as oily droplets very often causes the coagulation of catalyst particles, resulting in the termination of the reaction. Therefore, this must be avoided either by lowering the reaction temperature or by adding a suitable organic cosolvent that causes no clotting of the catalyst. Coagulation of the catalyst can also be produced by residues of some organic solvents. Solvent combinations employed in Heyns oxidation include acetone and water[12]; dioxane and water[13]; EtOAc and water[14]; diglyme and water[15]; isopropyl alcohol, EtOAc, and water[16]; and isopropyl alcohol, acetone, and water.[17]
A trace of a surfactant such as Dow Corning Antifoam A[18] or sodium lauryl sulfate[15] is sometimes added to facilitate the solubilization of the alcohol.

[k] The speed for the oxidation of primary alcohols increases with the pH and it is very low under acidic conditions. As the oxidation advances, the generation of a carboxylic acid would produce an acidic pH that would eventually cause the reaction to halt before conclusion. This is avoided by the addition of base. When just 1 equivalent of base is added, there is a decrease of the pH—and of oxidation velocity—as the carboxylic acid is generated that helps to avoid overoxidations at the cost of a lower overall oxidation rate. The base is often added sequentially as the reaction advances in order to maintain a pH of ca. 7–9, which secures a suitable oxidation speed while base-sensitive functional groups may remain unchanged. It must be mentioned that when a selective oxidation of a primary alcohol in the presence of a secondary one is sought, better results are expected under basic pH, because while the oxidation of primary alcohols is greatly suppressed under acidic conditions this may not be the case with secondary alcohols.[28a]

[l] Other bases are occasionally employed, including NaOAc,[1a, 21] KOH,[1a] potassium oxalate,[1a] potassium oxalate plus $NaHCO_3$,[1a] phosphate buffer,[1a, 22] $NaHCO_3$ plus K_2CO_3 (pH 7.5–9.0),[9a] $KHCO_3$,[23] $BaCO_3$,[7h] K_2CO_3,[24] and $NaHCO_3$ plus Na_2CO_3 (pH 8.8).[25] Heyns oxidations of primary alcohols are sometimes carried out with no base added.[27b, 28a, 28b, 12c, 16]

[m] It normally takes between 1 and 48 hours.

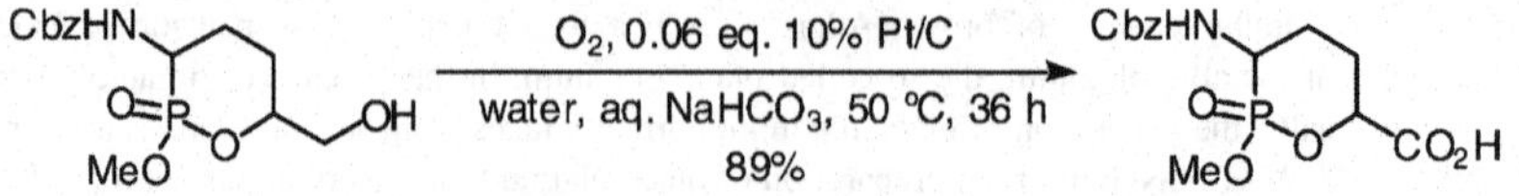
CbzHN
O=P
OH
MeO
O
O_2, 0.06 eq. 10% Pt/C
water, aq. $NaHCO_3$, 50 °C, 36 h
89%
CbzHN
O=P
CO_2H
MeO
O

Heyns oxidation secures 89% of the desired acid, while the reaction fails with chromium reagents, and Swern followed by permanganate gives a 40% yield.[42]

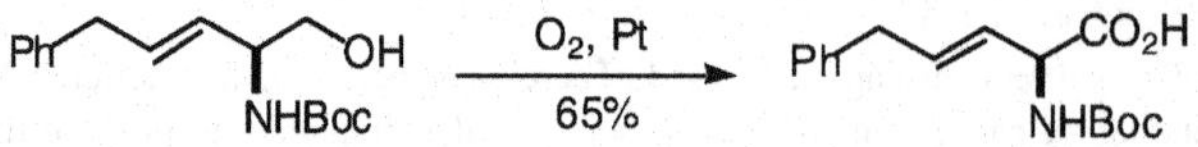
Ph
OH
NHBoc
O_2, Pt
65%
Ph
CO_2H
NHBoc

According to the authors[43]: "However, although the oxidation of analogous protected aminoalcohols by chemical reagents such as pyridinium dichromate or Jones' reagent has been reported, in our hands these methods were rather inefficient. We then attempted to exploit the oxidation catalyzed by molecular oxygen in the presence of platinum, and were pleased to find that this approach led to the acid that could be obtained in satisfactory yield (65%) and high optical purity."

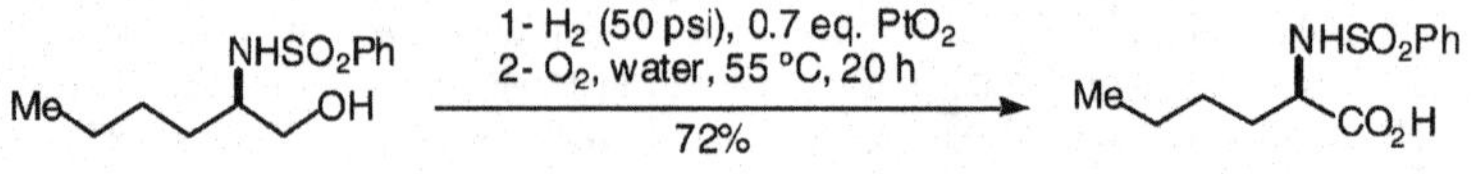
$NHSO_2Ph$
Me
OH
1- H_2 (50 psi), 0.7 eq. PtO_2
2- O_2, water, 55 °C, 20 h
72%
$NHSO_2Ph$
Me
CO_2H

While Heyns oxidation provides a 72% yield of the desired acid possessing optical purity above 99%, Jones oxidation, Sarett reagent, and PDC in DMF gave unsatisfactory results.[28a]

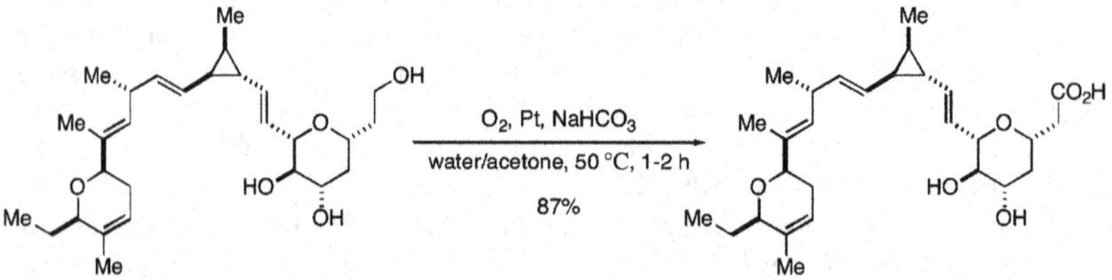
Me
Me
Me
O
Me
Me
OH
O
HO
OH
O_2, Pt, $NaHCO_3$
water/acetone, 50 °C, 1-2 h
87%
Me
Me
Me
O
Me
Me
CO_2H
O
HO
OH

In this complex and sensitive substrate it is possible to perform the selective oxidation of a primary alcohol in the presence of two secondary ones using platinum and oxygen.[17]

4.3. Functional Group and Protecting Group Sensitivity to Heyns Oxidation

Heyns oxidation normally transforms aldehydes into carboxylic acids[34, 44] under quite mild conditions, so that selective oxidation of aldehydes in the presence of primary alcohols is possible.[34, 44a, 45]

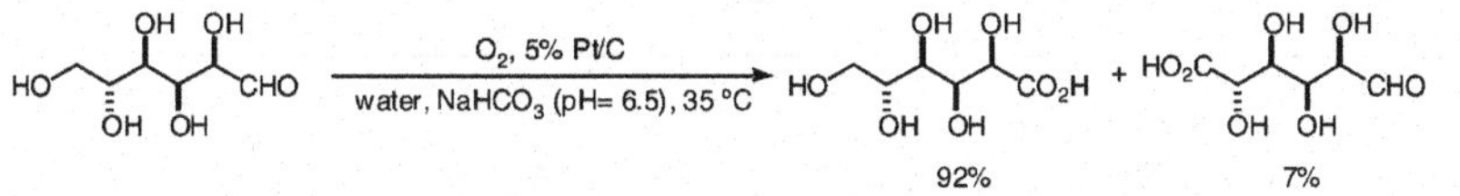

In the patent where this reaction is described, overoxidation is avoided by using an apparatus in which freshly formed monocarboxylic acids are continuously separated from the reaction medium by electrodialysis. Most of the aldehyde functionality (92%) is selectively oxidized in the presence of five alcohols, while minor amounts (7%) of product resulting from selective oxidation of the primary alcohol are also obtained.[46]

Secondary alcohols can be oxidized under Heyns' conditions.[2, 13a, 26a,c–f, 32a] Nevertheless, as the oxidation of primary alcohols can be made under milder conditions, Heyns oxidation is in fact one of the best methods to achieve selective oxidation of primary alcohols in the presence of secondary ones.[32a, 34, 38]

Heyns oxidation is normally carried out under basic conditions. This does not need to cause any interference with base-sensitive functional groups, because experimental conditions are normally quite mildly basic. In fact, oxidation of primary alcohols with platinum and oxygen can take place in the presence of base-sensitive functional groups such as esters,[16, 47] amides,[14, 48] urethanes,[42, 49, 43, 38a] and epoxides.[12d] In some cases a too basic pH causes interference with some base-sensitive functional groups.[47]

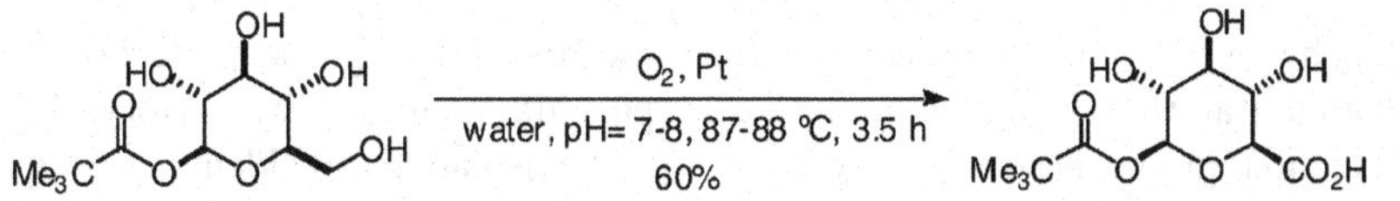

During the selective oxidation of a primary alcohol in the presence of three secondary ones, the pH must not be raised above 8 to avoid the hydrolysis of the base-sensitive ester.[47]

Although amines are reported to poison the platinum catalyst,[7j, 27c, 37] experimental conditions can be devised allowing the selective oxidation of primary alcohols in the presence of amines,[50] including pyridines.[12d]

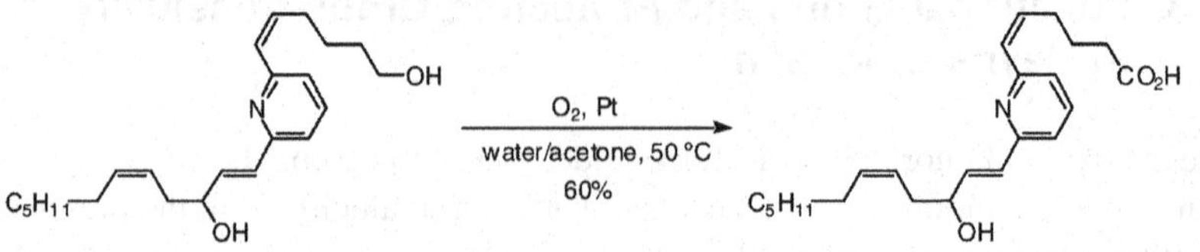

The oxidation of this primary alcohol using platinum and oxygen succeeds regardless
of the presence of a pyridine ring that could poison the
catalyst.[12d]

On the other hand, protection of amines as urethanes[38a, 42, 43, 49] or amides[14, 48] prevents any possible poisoning. *N*-Methylamines can be oxidized to a formamide[51] or demethylated[52] with platinum and oxygen.

Although low-valent sulfur compounds are known to poison the platinum catalyst,[7j, 36] there is one report in which an oxidation is made with platinum and oxygen in the presence of a thioacetal.[52] Sulfur compounds of high oxidation state are normally not able to coordinate with active sites in platinum and therefore cause no poisoning. Thus, oxidation of alcohols to carboxylic acids employing Heyns' method can be carried out, for example, in the presence of sulfones.[28b, 16]

Due to the mildness of Heyns' conditions, the oxidation of alcohols to acids can take place in the presence of many functional groups and moieties sensitive to oxidation, such as alkenes,[12b,d, 53, 43] phenols,[54] and aromatic compounds, including very oxidation-prone electron-rich aromatic compounds.[32b] Nonetheless, indoles, which are normally very sensitive to oxidation, are known to react with oxygen in the presence of platinum.[55]

4.4. Side Reactions

Aldehydes are sometimes isolated from the reaction of primary alcohols with platinum and oxygen.[2] This may be purposefully achieved by performing the oxidation in the absence of water.[11, 13a] The isolation of aldehydes is quite difficult in the presence of an excess of water, because they are very quickly oxidized to carboxylic acids via the corresponding hydrates.

During the oxidation of primary alcohols, secondary ones present in the same molecule can also suffer oxidation in a minor proportion.[34] When the oxidation of a secondary alcohol leads to an α-hydroxyketone, this is further oxidized resulting in a carbon–carbon bond breakage that many times results in generation of low-molecular-weight acids,[26f, 34] which may remain undetected.

A secondary alcohol such as *sec*-butanol[48a] or isopropyl alcohol[17] is sometimes added to the reaction medium, apparently with the purpose of mitigating the oxidation of secondary alcohols in the presence of primary ones.

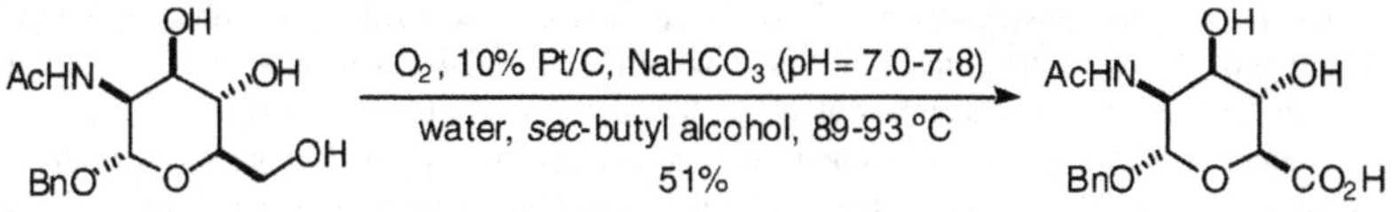

According to the authors: "over-oxidation was controlled by the addition of *sec*-butanol."[48a]

Sometimes, due to the basic medium, epimerization at the carbon adjacent to the carboxylic acid can occur.[56] This can be avoided by a proper choice of base.

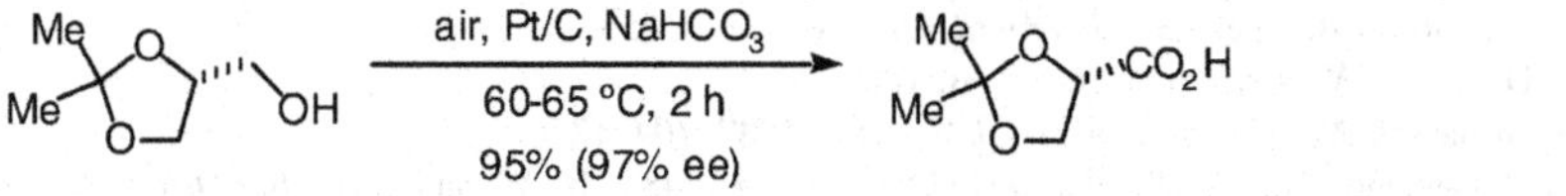

According to the authors: "Catalytic oxidation with Pt/C and air (or oxygen) in aqueous and basic medium gave excellent chemical yields of the acid. However, the choice of base was critical in avoiding any racemization of the chiral centre."[56]

The oxidation of 1,4- and 1,5-diols sometimes leads to formation of stable five- and six-membered lactones, instead of the corresponding hydroxyacids.[15, 57]

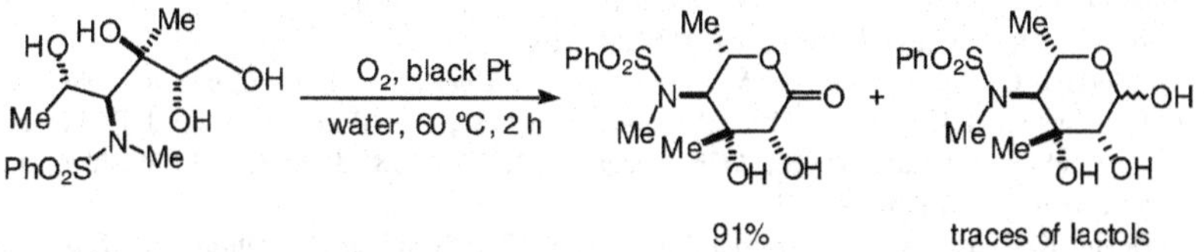

The treatment of the starting tetrol with oxygen in the presence of black platinum leads to the selective oxidation of the primary alcohol. A stable six-membered lactone is isolated instead of a free carboxylic acid. Traces of lactols in equilibrium with an aldehyde resulting from an incomplete oxidation are also obtained.[57a]

In this remarkable transformation, a five-membered lactone is quantitatively obtained by the selective oxidation of the primary alcohol. One of the secondary alcohols is transformed into a ketone, while the other secondary alcohol is protected *in situ* by the formation of a lactone. Notably, not even a trace of epimerization at the α position of the lactone is observed. According to the authors: "In the absence of the surfactant the reaction proceeded sluggishly, due to the insolubility of the starting alcohol, affording the product in ca. 50% yield after 40 h at 55 °C."[15]

4.5. References

1 a) Dalmer, O.; Heyns, K.; U.S. Pat. 2,190,377 **1940**. b) Heyns, K.; *Lieb.Ann.Chem.* **1947**, *558*, 177.

2 Heyns, K.; Paulsen, H.; *Angew.Chem.* **1957**, *69*, 600.

3 Strecker, A.; *Lieb.Ann.Chem.* **1855**, *93*, 370.

4 von Gorup-Besanez, E.; *Lieb.Ann.Chem.* **1861**, *118*, 259.

5 Dafert, F. W.; *Ber.Dtsch.Chem.Ges.* **1884**, *17*, 227.

6 Grimeaux, M.; *Compt.Rend.Hebd.Acad.Sci.* **1887**, *104*, 1276.

7 a) Wieland, H.; *Ber.Dtsch.Chem.Ges.* **1912**, *45*, 484. b) Wieland, H.; *Ber.Dtsch.Chem.Ges.* **1912**, *45*, 2606. c) Wieland, H.; *Ber.Dtsch.Chem.Ges.* **1913**, *46*, 3327. d) Wieland, H.; *Ber.Dtsch.Chem.Ges.* **1921**, *54*, 2353. e) Müller, E.; Schabe, K.; *Z.Elektrochem.* **1928**, *34*, 170. f) Müller, E.; Schabe, K.; *Kolloid Z.* **1930**, *52*, 163. g) Macrae, Th. F.; *Biochem.J.* **1933**, *27*, 1248. h) Rottenberg, M.; Baertschi, P.; *Helv.Chim.Acta* **1956**, *34*, 1973. i) Rottenberg, M.; Thürkauf, M.; *Helv.Chim.Acta* **1959**, *42*, 226. j) Heyns, K.; Paulsen, H.; *Newer Methods of Preparative Organic Chemistry*, Vol. 2, W. Forest, F. K. Kirchner, Ed.; Academic Press: New York, **1963**, p. 303–335. k) de Wit, G.; de Vlieger, J. J.; Kock-van Dalen, A. C.; Kieboom, A. P. G.; van Bekkum, H.; *Tetrahedron Lett.* **1978**, *15*, 1327. l) van Dam, H. E.; Kieboom, A. P. G.; van Bekkum, H.; *Appl.Catal.* **1987**, *33*, 361. m) Brönnimann, C.; Bodnar, Z.; Aeschimann, R.; Mallat, T.; Baiker, A.; *J.Catal.* **1996**, *161*, 720.

8 Weidmann, H.; Zimmerman Jr., H. K.; *Lieb.Ann.Chem.* **1961**, *639*, 198.

9 See, for example: a) Marsh, C. A.; *J.Chem.Soc.* **1952**, 1578. b) Tsou, K.-C.; Seligman, A. M.; *J.Am.Chem.Soc.* **1953**, *75*, 1042. c) Pradvić, N.; Keglević, D.; *Tetrahedron* **1965**, *21*, 1897.

10 Sneeden, R. P. A.; Turner, R. B.; *J.Am.Chem.Soc.* **1955**, *77*, 190.

11 Karrer, P.; Hess, W.; *Helv.Chim.Acta* **1957**, *40*, 265.

12 a) Moody, C. J.; Roberts, S. M.; Toczek, J.; *J.Chem.Soc. Chem.Commun.* **1986**, 1292. b) Hutchinson, D. K.; Fuchs, P. L.; *J.Am.Chem.Soc.* **1987**, *109*, 4755. c) Marino, J. P.; Fernández de la Pradilla, R.; Laborde, E.; *J.Org.Chem.* **1987**, *52*, 4898. d) Morris, J.; Wishka, D. G.; *Tetrahedron Lett.* **1988**, *29*, 143.

13 a) Heyns, K.; Blazejewicz, L.; *Tetrahedron* **1960**, *9*, 67. b) Sakaitani, M.; Ohfune, Y.; *Tetrahedron Lett.* **1987**, *28*, 3987.

14 Yabe, Y.; Guillaume, D.; Rich, D. H.; *J.Am.Chem.Soc.* **1988**, *110*, 4043.

15 Johnson, W. S.; Chan, M. F.; *J.Org.Chem.* **1985**, *50*, 2598.

16 Roemmele, R. C.; Rapoport, H.; *J.Org.Chem.* **1989**, *54*, 1866.

17 Liu, P.; Jacobsen, E. N.; *J.Am.Chem.Soc.* **2001**, *123*, 10772.

18 Bennani, Y. L.; Vanhessche, K. P. M.; Sharpless, K. B.; *Tetrahedron: Asymmetry* **1994**, *5,* 1473.

19 Gläser, R.; Josl, R.; Williardt, J.; *Top. Catal.* **2003,** *22,* 31.

20 Brönnimann, C.; Mallat, T.; Baiker, A.; *J.Chem.Soc. Chem.Commun.* **1995,** 1377.

21 Görlich, B.; German Pat. DE 935,968 **1955**.

22 Dettwiler, J. E.; Lubell, W. D.; *J.Org.Chem.* **2003,** *68,* 177.

23 a) Overend, W. G.; Shafizadeh, F.; Stacey, M.; Vaughan, G.; *J.Chem.Soc.* **1954,** 3633. b) Shao, Y.-Y.; Seib, P. A.; Kramer, K. J.; Van Galen, D. A.; *J.Agric. Food Chem.* **1993,** *41,* 1391.

24 Jacobson, B.; Davidson, E. A.; *Nature* **1961,** *189,* 663.

25 Moss, G. P.; Reese, C. B.; Schofield, K.; Shapiro, R.; Lord Todd; *J.Chem.Soc.* **1963,** 1149.

26 a) Heyns, K.; Paulsen, H.; *Chem.Ber.* **1953,** *86,* 833. b) Sneeden, R. P. A.; Turner, R. B.; *J.Am.Chem.Soc.* **1955,** *77,* 130. c) Heyns, K.; Paulsen, H.; *Chem.Ber.* **1956,** *89,* 1152. d) Eugster, C. H.; Waser, P. G.; *Helv.Chim.Acta* **1957,** *40,* 888. e) Heyns, K.; Gottschalck, H.; *Chem.Ber.* **1961,** *94,* 343. f) Heyns, K.; Lenz, J.; Paulsen, H.; *Chem.Ber.* **1962,** *95,* 2964.

27 a) Heyns, K.; Vogelsang, G.; *Chem.Ber.* **1954,** *87,* 13. b) Heyns, K.; Beck, M.; *Chem.Ber.* **1956,** *89,* 1648. c) Barker, S. A.; Bourne, E. J.; Fleetwood, J. G.; Stacey, M.; *J.Chem.Soc.* **1958,** 4128.

28 a) Maurer, P. J.; Takahata, H.; Rapoport, H.; *J.Am.Chem.Soc.* **1984,** *106,* 1095. b) Compagnone, R. S.; Rapoport, H.; *J.Org.Chem.* **1986,** *51,* 1713.

29 Paulsen, H.; Koebernick, W.; Autschbach, E.; *Chem.Ber.* **1972,** *105,* 1524.

30 van Dam, H. E.; Duijverman, P.; Kieboom, A. P. G.; van Bekkum, H.; *Appl.Catal.* **1987,** *33,* 373.

31 Glattfeld, J. W. E.; Gershon, S.; *J.Am.Chem.Soc.* **1938,** *60,* 2013.

32 a) van Dam, H. E.; Kieboom, A. P. G.; van Bekkum, H.; *Recl.Trav.Chim.Pays-Bas* **1989,** *108,* 404. b) Ainge, D.; Ennis, D.; Gidlund, M.; Stefinovic, M.; Vaz, L.-M.; *Org.Proc.Res.Dev.* **2003,** *7,,* 198.

33 Heyns, K.; Lenz, J.; *Chem.Ber.* **1961,** *94,* 348.

34 Venema, F. R.; Peters, J. A.; van Bekkum, H.; *J.Mol.Catal.* **1992,** *77,* 75.

35 Sidorov, S. N.; Volkov, I. V.; Davankov, V. A.; Tsyurupa, M. P.; Valetsky, P. M.; Bronstein, L. M.; Karlinsey, R.; Zwanziger, J. W.; Matveeva, V. G.; Sulman, E. M.; Lakina, N. V.; Wilder, E. A.; Spontak, R. J.; *J.Am.Chem.Soc.* **2001,** *123,* 10502.

36 Wacek, A.; Limontschew, W.; Leitinger, F.; Hilbert, F.; Oberbichler, W.; *Monatsh.Chem.* **1959,** *90,* 555.

37 Trenner, N. R.; U.S. Pat. 2,428,438 **1947**.

38 See, for example: a) Hasuoka, A.; Nishikimi, Y.; Nakayama, Y.; Kamiyama, K.; Nakao, M.; Miygawa, K.-I.; Nishimura, O.; Fujno, M.; *J.Antibiot.* **2002,** *55,* 191. b) Fabre, J.; Betbeder, D.; Paul, F.; Monsan, P.; *Synth.Commun.* **1993,** *23,* 1357.

39 Tsuchioka, T.; Yamaguchi, T.; Yuuen, K.; Chaen, H.; Eur.Pat.Appl. 864,580 **1998**.

40 a) Edye, L. A.; Meehan, G. V.; Richards, G. N.; *J.Carbohydr.Chem.* **1991,** *10,* 11. b) Johnson, L.; Verraest, D. L.; van Haveren, J.; Hakala, K.; Peters, J. A.; van Bekkum, H.; *Tetrahedron: Asymmetry* **1994,** *5,* 2475.

41 Adams, R.; Voorhees, V.; Shriner, R. L.; *Org.Synth.Coll. I* 2nd ed., 463.

42 Bartlett, P. A.; Vanmaele, L. J.; Kezer, W. B.; *Bull.Soc.Chim.Fr.* **1986,** 776.

43 Mehmandoust, M.; Petit, Y.; Larchevêque, M.; *Tetrahedron Lett.* **1992,** *33,* 4313.

44 a) Smits, P. C. C.; Kuster, B. F. M.; Van der Wiele, K.; Van der Baan, H. S.; *Carbohydr.Res.* **1986,** *153,* 227. b) Verdeguer, P.; Merat, N.; Gaset, A.; *J.Mol.Catal.* **1993,** *85,* 327.

45 Abbadi, A.; Gotlieb, K. F.; Meiberg, J. B. M.; van Bekkum, H.; *Appl.Catal. A: General* **1997,** *156,* 105.

46 Markwart, K.; German Pat. DE 4,307,388 **1994**.

47 Bols, M.; *J.Org.Chem.* **1991,** *56,* 5943.

48 a) Yoshimura, J.; Sakai, H.; Oda, N.; Hashimoto, H.; *Bull.Chem.Soc.Jpn.* **1972,** *45,* 2027. b) Jeffs, P. W.; Chan, G.; Sitrin, R.; Holder, N.; Roberts, G. D.; DeBrosse, C.; *J.Org.Chem.* **1985,** *50,* 1726.

49 Kogen, H.; Nishi, T.; *J.Chem.Soc. Chem.Commun.* **1987,** 311.

50 a) Blaufelder, C.; Broucek, R.; Carstens, A.; Eisenhuth, L.; PCT Pat.Appl. WO 2001/010818 **2001**. b) Kimura, H.; Japan Pat. 5,140,056 **1993**.

51 Davis, G. T.; Rosenblatt, D. H.; *Tetrahedron Lett.* **1968,** 4085.

52 Birkenmeyer, R. D.; Dolak, L. A.; *Tetrahedron Lett.* **1970,** 5049.

53 Casy, G.; Taylor, R. J. K.; *J.Chem.Soc. Chem.Commun.* **1988,** 454.

54 Moryasu, M.; Maki, T.; Araki, Y.; Japan Pat. 5,194,309 **1993.**

55 Harley-Mason, J.; Taylor, C. G.; *J.Chem.Soc. Chem.Commun.* **1970,** 812.

56 Handa, S.; Hawes, J. E.; Pryce, R. J.; *Synth.Commun.* **1995,** *25,* 2837.

57 a) Maurer, P. J.; Knudsen, C. G.; Palkowitz, A. D.; Rapoport, H.; *J.Org.Chem.* **1985,** *50,* 325.
    b) González-De-La-Parra, M.; Hutchinson, C. R.; *J.Am.Chem.Soc.* **1986,** *108,* 2448.

5

Ruthenium Tetroxide and Other Ruthenium Compounds

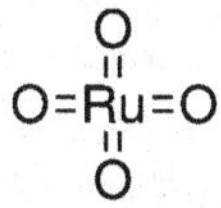

$$O=\overset{\displaystyle O}{\underset{\displaystyle O}{\overset{\|}{\underset{\|}{Ru}}}}=O$$

5.1. Introduction

Ruthenium tetroxide (RuO_4) is a golden-yellow volatile solid with an acrid odor, sparingly soluble in water and freely soluble in CCl_4, in which it forms stable solutions. In comparison with the analogous compound OsO_4, ruthenium tetroxide is a stronger oxidizing agent that reacts violently—resulting in explosion and/or flames—with most common organic solvents such as ether, alcohols, benzene, and pyridine,[1] and also with filter paper.

> Ruthenium tetroxide sublimes very easily at room temperature, possessing a melting point of 25.4 °C and a boiling point of 40 °C. While it is very soluble in CCl_4 and other nonflammable organic solvents, a saturated solution in water at 20 °C reaches a concentration of 2% w/v.

WARNING: RuO_4 is a toxic and explosive compound, and, although it is less toxic than OsO_4, it must be handled in a well ventilated fume hood using goggles and gloves. It can be destroyed with a sodium bisulfite solution, resulting in the much safer and less toxic ruthenium dioxide, which is a dark insoluble solid with very low vapor pressure.

Ruthenium tetroxide can be conveniently handled as a carbon tetrachloride solution that is easily prepared by stirring an aqueous solution of sodium periodate ($NaIO_4$) with a suspension of hydrated ruthenium dioxide in CCl_4.[2] Ruthenium tetroxide partitions between CCl_4 and water, resulting in a 60:1 concentration ratio.[3]

Procedure of Pappo and Becker

In 1953, Djerassi and Engle[1] reported the reaction of stoichiometric RuO_4 in CCl_4 with a number of organic compounds. Three years later, Pappo and Becker published in a journal of limited distribution[4] the use of catalytic RuO_4 in the presence of $NaIO_4$ as secondary oxidant, in the oxidation of alkenes and alkynes. In 1958, Berkowitz and Rylander[5] described the oxidation of *n*-hexanol to hexanoic acid using a *stoichiometric* solution of RuO_4 in CCl_4.

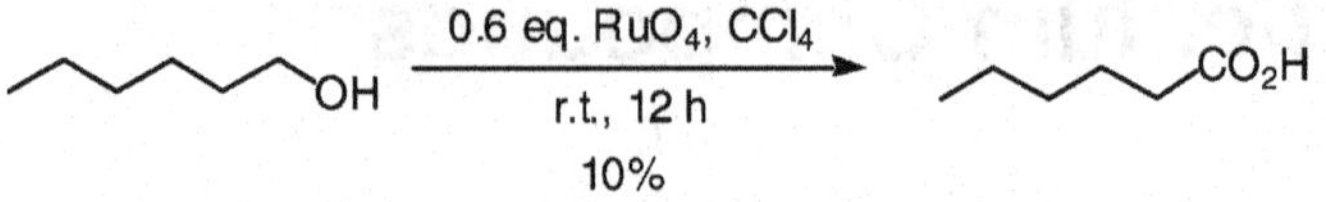

The first oxidation of a primary alcohol to carboxylic acid using RuO_4 was reported by Berkowitz and Rylander in 1958. They obtained a modest 10% yield of hexanoic acid by treating *n*-hexanol with a stoichiometric quantity of RuO_4 dissolved in CCl_4.[5]

In 1969, during the synthesis of a mold metabolite, Roberts *et al.*[6] employed for the first time *catalytic* RuO_4 in the oxidation of primary alcohols to carboxylic acids. The use of catalytic RuO_4, normally in the presence of $NaIO_4$ as secondary oxidant, became very quickly routine in the oxidation of organic compounds with RuO_4 because of economy and safety.

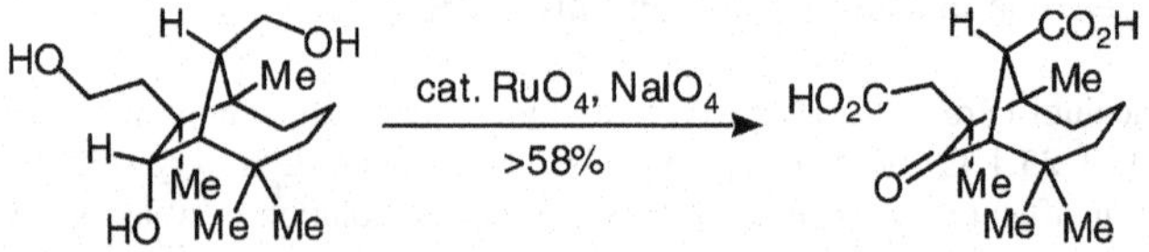

The first use of catalytic RuO_4 for the oxidation of primary alcohols to carboxylic acids was reported by Roberts *et al.* in 1969 during the preparation of the mold metabolite culmorin.[6] They employed the catalytic RuO_4 oxidation procedure of Pappo and Becker[4]—consisting of the use of a small amount of RuO_4, generated from RuO_2, in the presence of excess of $NaIO_4$— to achieve the simultaneous oxidation of two primary alcohols to carboxylic acids and one secondary alcohol to ketone.

Sharpless' Modification

In 1981, Sharpless *et al.* noted in a repeatedly cited paper[7] the eventual inactivation of the RuO_4 catalytic cycle in oxidations in which carboxylic acids are produced. This was attributed to formation of complexes between carboxylic acids and low-valent ruthenium compounds, in which ruthenium resisted the reoxidation to RuO_4. They found that acetonitrile prevents the inactivation of the RuO_4 catalytic cycle, something that they attributed to acetonitrile competing efficiently with carboxylic acids for complexation with low-valent ruthenium compounds, resulting in complexes in which the reoxidation to RuO_4 can be effective. They proposed the use of the solvent mixture water/MeCN/CCl_4 in a 3:2:2 ratio for oxidations carried out using the procedure of Pappo and Becker with catalytic RuO_4 and excess of $NaIO_4$. Although they reported in the first publication a single example of oxidation of a primary alcohol to carboxylic acid employing their modified protocol, Sharpless' modification of the procedure of Pappo and Becker found very quickly a widespread use in the preparation of carboxylic acids from primary alcohols because of its consistent efficiency and very easy experimental conditions.

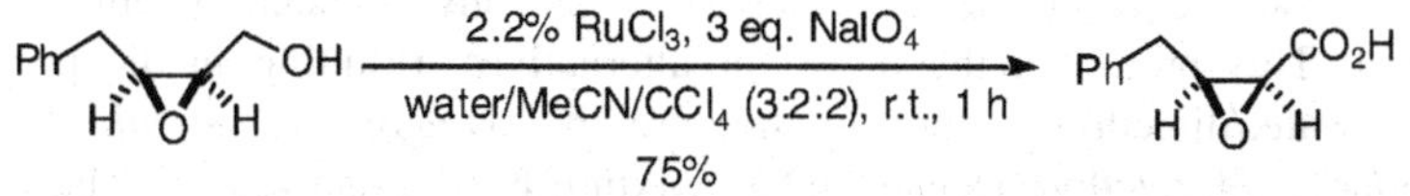

After observing that, in oxidations in which carboxylic acids are produced, the catalytic effect of RuO_4 ceases due to complexation of carboxylic acids with low-valent ruthenium compounds, Sharpless *et al.* proposed the use of the solvent mixture water/MeCN/CCl$_4$ in a 3:2:2 ratio, which gives optimum results in the oxidation of (E)-5-decene to pentanoic acid.[7] Although this oxidation protocol was not optimized for the oxidation of primary alcohols to carboxylic acids, and Sharpless *et al.* included in their first paper only the above example of this functional group transformation, Sharpless' modification of the protocol of Pappo and Becker[4] for oxidations with catalytic RuO_4 found very quickly extensive use in the preparation of carboxylic acids from primary alcohols.

Mechanism

The rate-determining step in the oxidation of alcohols with RuO_4 consists of a hydride transfer from the alcohol to RuO_4, leading to a protonated carbonyl group.[8] The experimental data are consistent with the mechanism shown below for the obtention of carboxylic acids, in which RuO_4 oxidizes the alcohol to an aldehyde, that is hydrated to a *gem*-diol before a second oxidation delivering a carboxylic acid occurs.

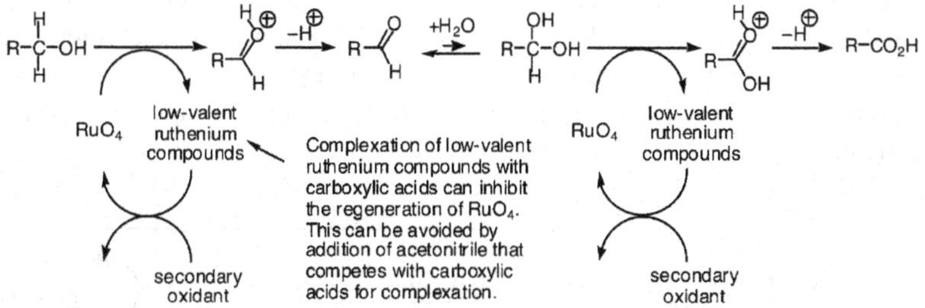

Solvent

In the vast majority of cases, Sharpless' modification of the procedure of Pappo and Becker for catalytic oxidations with RuO_4 is followed, and a solvent mixture consisting of water, acetonitrile, and CCl$_4$ in a 3:2:2 ratio is employed.

Slight variations of the water/MeCN/CCl$_4$ (3:2:2) ratio are sometimes used to fit diverse solubilities of substrate, resulting in successful oxidations.[9] Thus, the proportion of water is sometimes decreased to a ratio between 3:2:2 and 1.5:2:2. Sometimes, CH$_2$Cl$_2$[10] or CHCl$_3$[11] is used in the place of CCl$_4$. On occasions, no CCl$_4$ is added and a mixture of water and acetonitrile, containing a water/MeCN ratio ranging from 1:1[12] to 1:6,[13] is employed. It must be mentioned that the use of acetonitrile as the main solvent, with the addition of a few equivalents of water, promotes the isolation of the intermediate aldehyde.[14]

Carbon tetrachloride is a poisonous and environmentally objectionable solvent. This prompted the quest for alternatives, resulting in the proposals of a water/dimethyl carbonate mixture by Dragojlovic *et al.*[15] and a water/acetonitrile/ethyl acetate (3:2:2) mixture by Prashad *et al.*[16] The solvent choice is severely limited by the strong oxidizing power of RuO$_4$.

The oxidation of phenethyl alcohol using the water/MeCN/CCl$_4$ (3:2:2) solvent mixture proposed by Sharpless *et al.* results in an 80% yield of phenylacetyl acid. The substitution of CCl$_4$ for EtOAc, which is less toxic and less damaging to the environment, following the recommendation of Prashad *et al.*, allows the obtention of an improved 93% yield of carboxylic acid. Alternatively, the more expensive solvent PhCF$_3$ can be employed in place of CCl$_4$, resulting in 90% yield of carboxylic acid.[16]

A water/acetone mixture is very often used, resulting in successful oxidations with no apparent inactivation of the RuO$_4$ cycle.[17] Normally, a water/acetone ratio between 2:1[17a] and 1:5[18] is used. The inactivation of the RuO$_4$ cycle by carboxylates, first described by Sharpless *et al.*, was later confirmed by Boelrijk and Reedijk, who conducted some very convincing experiments.[8c] The successful implementation of oxidations, with no acetonitrile being added to prevent the inactivation of the RuO$_4$ cycle, proves that either not all carboxylates deactivate the RuO$_4$ cycle, or acetone is able to prevent the inactivation.

The alcohol is successfully oxidized with no inactivation of the RuO$_4$ catalytic cycle despite the absence of acetonitrile. According to the authors "We did not experience any difficulty with the oxidation, performed here in aqueous acetone, due to the formation of (inactive) lower valent ruthenium carboxylates."[19]

Buffering

A buffer phosphate solution is sometimes employed instead of plain water, in order to avoid side reactions.[20]

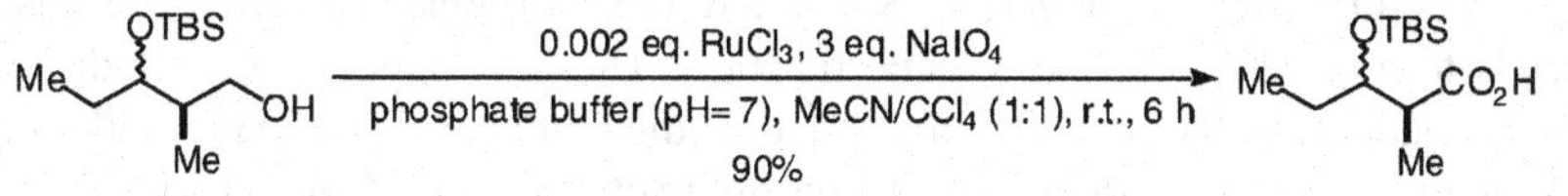

This alcohol oxidation is done in the presence of neutral phosphate buffer to prevent cleavage of the TBS ether.[20a]

Some equivalents of $NaHCO_3$,[21] or less commonly K_2CO_3,[11b] are often added, because this leads to the stabilization of the acid as the sodium (or potassium) carboxylate.

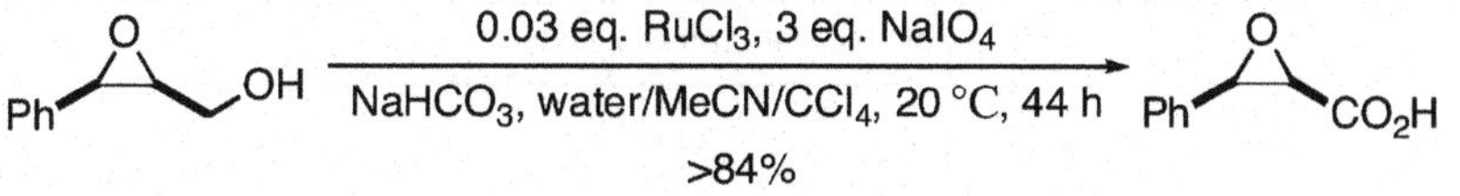

The yield of this oxidation is much higher in the presence of a few equivalents of $NaHCO_3$ that serve to stabilize the acid, as it is formed, as its sodium salt.[22]

Interestingly, oxidations with catalytic RuO_4 can also be carried out under slightly acidic conditions using an $EtOAc/water/CF_3CO_2H$ mixture containing a small proportion of CF_3CO_2H.[23]

RuO_4 Source

In the majority of cases, $RuCl_3$—normally in hydrated form— is employed as the source of catalytic RuO_4. Hydrated RuO_2 is less often used.

During the generation of *stoichiometric* RuO_4, it was observed that hydrated RuO_2 from different vendors possesses diverse reactivity, RuO_2 with higher content of water being the one with the greatest reactivity.[8a, 24] No such diverse reactivities of hydrated RuO_2 have been reported during the generation of catalytic RuO_4, although such outcome would not be unexpected.

During the oxidation of some sensitive alcohols, Falorni *et al.* reported that the use of RuO_2 instead of $RuCl_3$ caused oxidative decomposition of the substrates.[25]

Sabbatini *et al.* prepared a form of RuO_2 deposited on Teflon[26] that can be employed as a very convenient recyclable catalyst in the oxidation of alcohols.[27]

Secondary Oxidant

In the vast majority of cases sodium periodate ($NaIO_4$) is used as secondary oxidant in the transformation of primary alcohols into carboxylic acids employing catalytic RuO_4. Potassium periodate is also effective.[17a,b,d]

On the other hand, Chong and Sharpless recommend[28] the use of periodic acid (H_5IO_6), first employed by Stock and Tse,[29] in place of $NaIO_4$, because it leads to a quicker and more facile reaction. The effectiveness of periodic acid as secondary oxidant was confirmed by other authors.[30] Periodic acid is not compatible with substrates that exhibit high sensitivity to acidic conditions.[30c]

According to the authors "We have recently found that the use of periodic acid instead of sodium periodate in this procedure provides for a more facile reaction. Typically, use of 2.5 equivalents of H_5IO_6 and 2 mol% $RuCl_3$ in the described CCl_4-CH_3CN-H_2O system effects complete conversion to the acid (with no trace of intermediate aldehyde) within 2 h at room temperature. The superiority of periodic acid over sodium periodate in the acetonitrile-modified RuO_4 oxidations of aromatic compounds was brought to our attention by Mr. Kwok Tse and Professor Leon Stock."[28]

Other secondary oxidants occasionally mentioned in the oxidation of alcohols to carboxylic acids with catalytic RuO_4 include hydrogen peroxide,[31] sodium bromate ($NaBrO_3$),[8c] bleach ($NaOCl$)[15] and Oxone® (potassium peroxymonosulfate).[15]

Ruthenate and Perruthenate Oxidations

The ruthenate (RuO_4^{2-}) and perruthenate (RuO_4^-) anions are reagents in which ruthenium exists in lower oxidation state than in RuO_4. As expected, they behave as milder oxidants than RuO_4 and find occasional use in oxidation of primary alcohols to carboxylic acids in sensitive substrates.

In 1972, Lee, Hall, and Cleland described[32] the oxidation of primary alcohols to carboxylic acids employing *stoichiometric* sodium ruthenate in basic aqueous solution. In 1979, Griffith and Schröder reported[33] the use of *catalytic* potassium ruthenate in the presence of potassium persulfate ($K_2S_2O_8$) as secondary oxidant. The catalytic procedure was rediscovered by Varma and Hogan in 1992,[34] and other authors[35] confirmed its efficiency in sensitive substrates. In contrast to RuO_4, potassium ruthenate is compatible with functionalities like alkenes, which react readily with RuO_4.

Oxidations of primary alcohols to carboxylic acids with potassium ruthenate are normally carried out by stirring at room temperature a solution of the alcohol in ca. 0.2–1 M aqueous KOH, containing ca. 0.007–0.02 equivalents of $RuCl_3$ and ca. 3–5.3 equivalents of potassium persulfate. It is necessary to use basic pH to guarantee the stability of the ruthenate anion.[8c, 32]

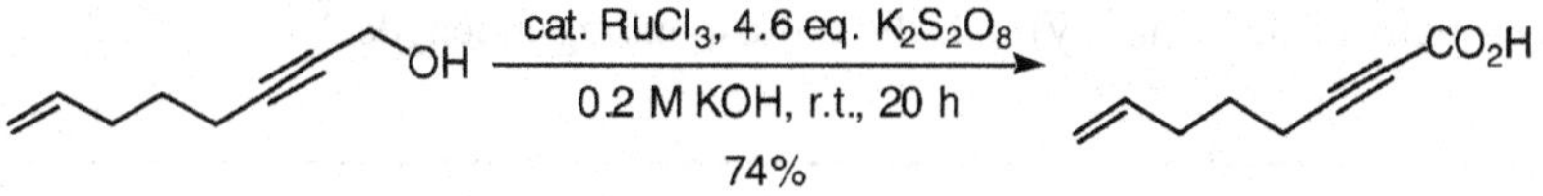

An alcohol is efficiently oxidized to carboxylic acid[35b] in the presence of an alkene and an alkyne, employing potassium ruthenate—a milder oxidant than RuO_4—generated by oxidation of catalytic $RuCl_3$ with potassium persulfate, using the method of Griffith and Schröder.

Research needed

An efficient method for oxidation of primary alcohols to carboxylic acids, employing catalytic ruthenate in an organic solvent under neutral conditions, must be developed. It would serve as a milder alternative to the use of catalytic RuO_4.

The perruthenate anion in the form of its tetra-*n*-propylammonium salt, the so-called TPAP (n-Pr_4NRuO_4), is a very common oxidant for the transformation of alcohols into aldehydes and ketones, in which case it is used in catalytic amounts and in the presence of N-methylmorpholine N-oxide (NMO) as secondary oxidant.[36] When this oxidation was first developed by Ley, Griffith *et al.*, they focused on the obtention of aldehydes and ketones. Therefore, they recommended the addition of molecular sieves to remove the water, which otherwise could promote overoxidation of primary alcohols to acids. On the other hand, when TPAP is used in the absence of molecular sieves over an extended time, it is many times possible to isolate the corresponding acid. The water necessary for the transformation of the aldehyde into carboxylic acid may have originated from the oxidation of the primary alcohol to the aldehyde, from the NMO secondary oxidant, which is normally sold hydrated, or may enter adventitiously into the reaction medium.

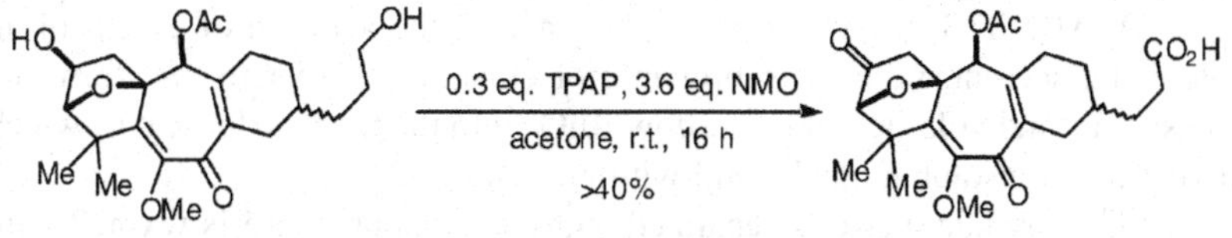

According to the authors: "surprisingly, the TPAP oxidation did not stop at the aldehyde stage but went on to give the corresponding acid."[37] Observe that additionally a secondary alcohol was oxidized to ketone.

Two protocols can be utilized when TPAP is employed for the transformation of primary alcohols into carboxylic acids. The reaction can be performed simply in the absence of molecular sieves over a prolonged time.[37, 38] Alternatively, a few equivalents of water can be added once the primary alcohol has been transformed (mainly) into the corresponding aldehyde.[39]

Research needed

TPAP is very rarely used for the oxidation of primary alcohols to carboxylic acids. TPAP is a very mild oxidant compatible with many functional groups. A detailed study of the potential and scope of catalytic TPAP for the oxidation of alcohols to carboxylic acids must be carried out.

Application

RuO_4 is a very strong oxidant that transforms very efficiently primary alcohols into carboxylic acids in robust substrates. It is particularly useful during oxidations in which the stereochemical integrity of the carbon atom being oxidized must be maintained. Due to the strong oxidizing power of RuO_4, this reagent is not suitable for substrates possessing functional groups, other than primary alcohols, sensitive to oxidation.

5.2. General Procedure for Oxidation of Primary Alcohols to Carboxylic Acids with Catalytic RuO_4

WARNING: ruthenium tetroxide, which is a very volatile (b.p. 40 °C) and poisonous compound, is generated during the reaction. Therefore, the reaction must be performed in a well-ventilated fume hood using gloves.

A biphasic mixture is prepared by dissolving 1 equivalent of the alcohol in a 3:2:2[a] mixture of water[b], acetonitrile,[c] and carbon tetrachloride.[d] This mixture must contain approximately from 8 to 56 g—typically 14 g—of alcohol per liter. Approximately, from 0.02 to 0.15 equivalent of $RuCl_3$[e] and ca. 2.25–6 equivalents of $NaIO_4$[f] (MW 213.89) are added, and the resulting mixture is vigorously stirred[g] untill most of the starting alcohol is consumed.[h]

WARNING: the workup must begin with the addition of excess of an alcoholic solvent, such as isopropanol or ethanol, in order to transform the poisonous and volatile golden-yellow RuO_4 into the much safer RuO_2, which exists as an insoluble black solid with no volatility.

The organic phase is separated and the aqueous phase is washed with an organic solvent such as CH_2Cl_2 or EtOAc. Optionally, some organic solvent such as CH_2Cl_2, Et_2O, or EtOAc and/or some water can be added to facilitate the separation of the organic phase. If the reaction is performed in the presence of NaHCO$_3$ or the resulting acid is very soluble in water, it is necessary to acidify the aqueous phase before separating the organic one. The precipitated RuO$_2$ can be optionally filtered. The collected organic phases are dried (Na$_2$SO$_4$ or MgSO$_4$) and concentrated, yielding a crude acid that may need further purification. Optionally, some degree of purification of the acid can be attained by extracting the collected organic phases with a basic aqueous solution such as 1 M NaOH or 5% NaHCO$_3$, and acidifying the basic aqueous phase. This sometimes leads to the crystallization or precipitation of the acid that can be separated by filtration. Alternatively, the acidified aqueous phase can be extracted with an organic solvent such as EtOAc, CH$_2$Cl$_2$ or Et$_2$O, and the organic phase can be dried (Na$_2$SO$_4$ or MgSO$_4$) and concentrated giving a crude acid that may need further purification.

[a] Slight variations of this volume ratio can be employed to accommodate the solubility of the alcohol.[9] Thus, the proportion of water can be decreased to 1.5:2:2. Sometimes no CCl$_4$ is added and a simple mixture of water and acetonitrile in a ratio from 1:1[12] to 1:6[13] is utilized.

[b] A neutral phosphate buffer is sometimes used instead of plain water to prevent alteration of sensitive functional groups.[20]
Alternatively, ca. 6.5–7 equivalents of NaHCO$_3$[21] are very often added to favor the oxidation by the stabilization of the resulting acid as the sodium carboxylate. K$_2$CO$_3$ is less often added.[11b]

[c] Acetonitrile must be added in order to prevent the inactivation of the RuO$_4$ catalytic cycle by complexation of the generated carboxylic acids with low-valent ruthenium compounds. Nevertheless, the oxidation can sometimes be successfully performed in a mixture of water and acetone in a 2:1[17a] to 1:5[18] ratio. A mixture of water and methyl carbonate, in which inactivation of the RuO$_4$ cycle does not occur, can also be employed.[15]

[d] Because of environmental concerns, ethyl acetate[16] is sometimes utilized rather than CCl$_4$. Less common solvents used in place of CCl$_4$ include CH$_2$Cl$_2$[10] and CHCl$_3$.[11]

[e] Normally, RuCl$_3$ is sold as a hydrate. Hydrated RuO$_2$ can be used in place of RuCl$_3$. Although RuCl$_3$ and RuO$_2$ are normally equally effective in the generation of RuO$_4$, some authors reported different behavior of these two reagents.[25]

[f] KIO$_4$ is equally efficient.[17a,b,d]

[g] The reaction is normally performed at room temperature. It can be carried out at 0 °C for milder conditions. Mixing of reagents is sometimes done at low temperature and the reaction mixture is left to reach (slowly) room temperature. Attempting to accelerate the reaction by heating can be dangerous and counterproductive due to the poisonous and highly volatile nature of RuO$_4$.

[h] It normally takes between 1 and 12 hours—typically 2 hours.

This oxidation, which fails using potassium permanganate, can be efficiently carried out with catalytic RuO$_4$.[40]

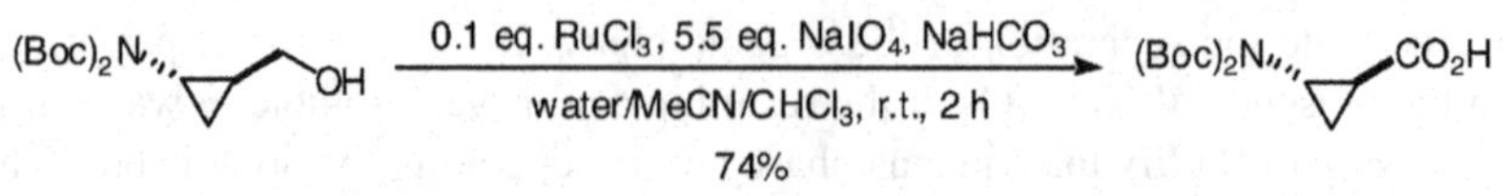

$$(Boc)_2N_{\prime\prime\prime}\text{—}\triangle\text{—}OH \xrightarrow[\substack{water/MeCN/CHCl_3,\ r.t.,\ 2\ h \\ 74\%}]{0.1\ eq.\ RuCl_3,\ 5.5\ eq.\ NaIO_4,\ NaHCO_3} (Boc)_2N_{\prime\prime\prime}\text{—}\triangle\text{—}CO_2H$$

This oxidation could not be satisfactorily performed with Jones reagent or with PDC, probably due to opening of the cyclopropane ring. However, the acid could be cleanly produced with catalytic RuO$_4$.[21e]

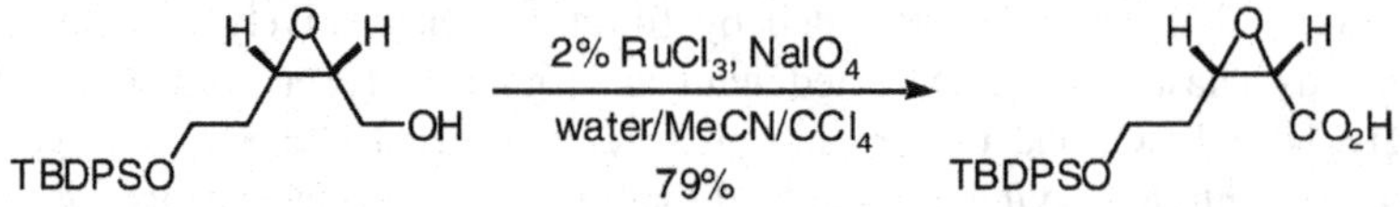

According to the authors "The oxidation of the epoxy alcohol was not a trivial matter. While many oxidants afforded only traces of carboxylic acid, pyridinium dichromate in wet DMF produced a 30–50% yield of the desired epoxy acid and Jones' reagent gave up to 60% epoxy acid but only with substantial epimerization to the trans isomer. It was found however that the new ruthenium-catalyzed oxidation of Sharpless and co-workers led cleanly to [the acid] in 79% yield without detectable epimerization."[41]

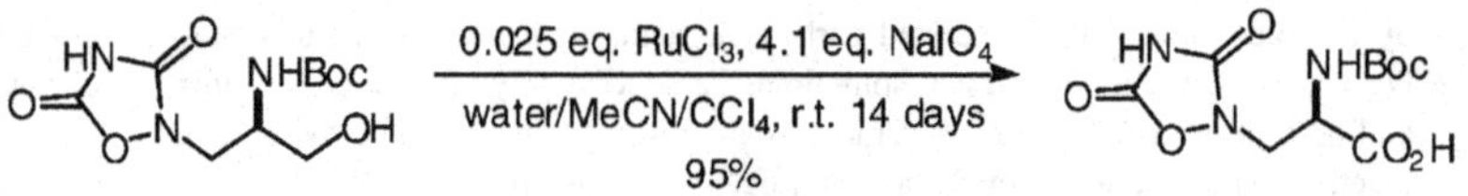

This transformation, which fails employing Jones reagent, PDC, or KMnO$_4$, can be carried out in a very good yield utilizing catalytic RuO$_4$.[42]

5.3. Functional Group and Protecting Group Sensitivity to Oxidation with Catalytic RuO$_4$

Ruthenium tetroxide is a very strong oxidant able to react with many functional groups including aldehydes[5], amines,[5] alkenes and alkynes,[5, 43] sulfides,[1] aromatic rings,[44] and oximes.[45] It is also capable of reacting with groups that are normally difficult to oxidize such as amides,[5] ethers,[5, 46] and even unfunctionalized alkanes.[47] Fortunately, many of these groups are oxidized under relatively harsh conditions and selective oxidations of primary alcohols with RuO$_4$ are possible. Thus, although normally it is not possible to oxidize primary alcohols in the presence of amines,[48] selective oxidations of alcohols are possible with amines protected as urethanes[10, 49] or amides.[50]

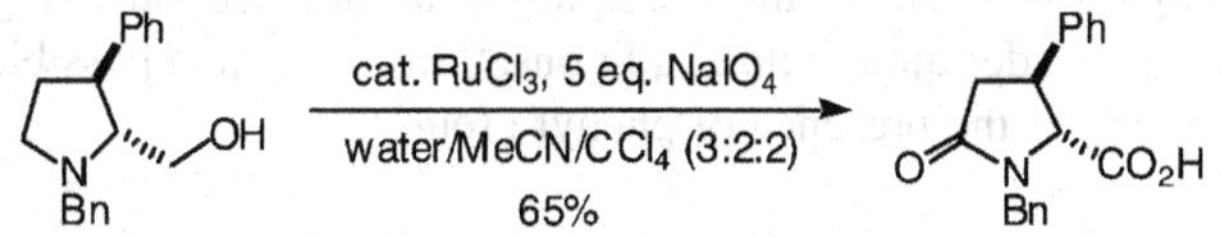

Under the action of catalytic RuO$_4$, a primary alcohol is transformed into a carboxylic acid and a tertiary amine into an amide. [48]

Alkenes normally suffer oxidative breakage producing two carboxylic acids under the reaction conditions used to oxidize primary alcohols, something that can sometimes be employed for synthetic advantage when both transformations are desired. [51]

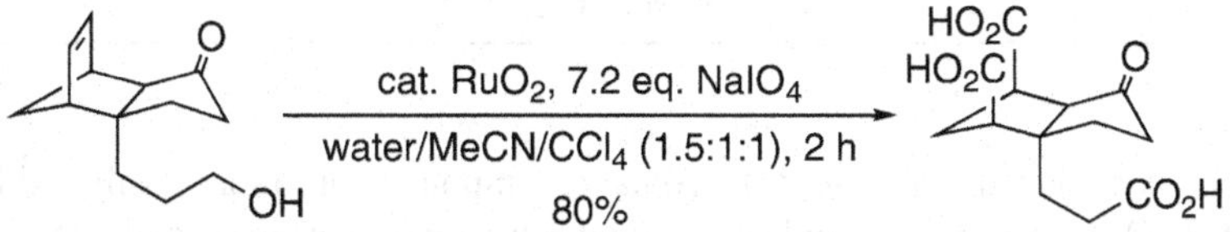

Catalytic RuO$_4$ is able concurrently to transform a primary alcohol into a carboxylic acid and to produce the oxidative breakage of an alkene to two carboxylic acids. [51]

On the other hand, it is sometimes possible to oxidize a primary alcohol in the presence of electron-deficient and hindered alkenes, which are less prone to suffer oxidation. [23]

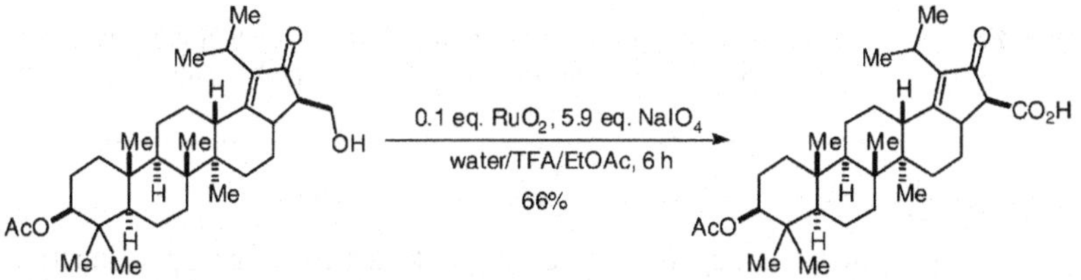

A primary alcohol is oxidized with catalytic RuO$_4$ in the presence of an alkene with a low reactivity to oxidation due to steric hindrance and electron deficiency, produced by conjugation with a carbonyl group. [23]

It must be mentioned that it is possible to oxidize primary alcohols in the presence of normal alkenes utilizing the ruthenate anion (see page 67), which possesses a lesser oxidation power than RuO$_4$.

Although RuO$_4$ is routinely employed to degrade phenyl groups into carboxylic acids under quite mild conditions,[44] normally it is possible to oxidize primary alcohols in the presence of phenyl groups.[35c, 52]

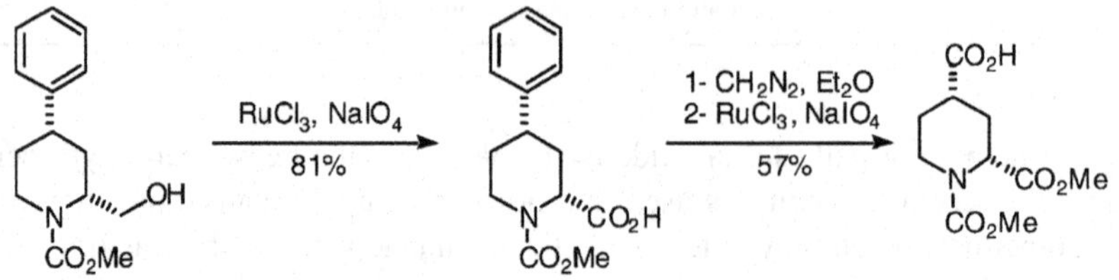

In this interesting example, a primary alcohol is selectively oxidized to a carboxylic acid with catalytic RuO$_4$ in the presence of a phenyl group, which later in the synthetic route is degraded to carboxylic acid using also catalytic RuO$_4$.[52d]

As expected, *electron-rich* aromatic compounds are more easily oxidized with RuO$_4$ than simple phenyl groups, and sometimes interfere with the oxidation of primary alcohols.[53] Nonetheless, sometimes it is possible to oxidize selectively primary alcohols in the presence of electron-rich aromatics.[54]

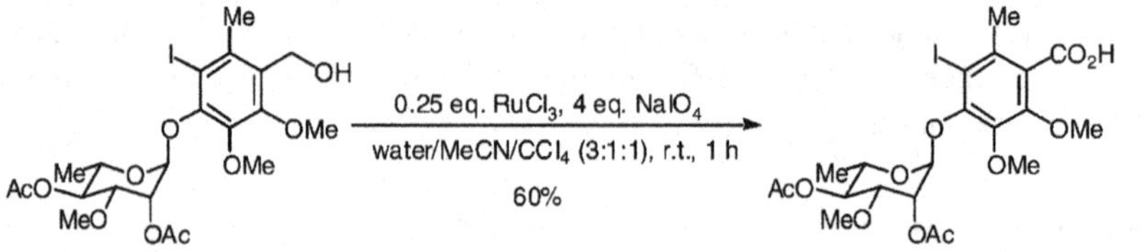

A very hindered primary alcohol is selectively oxidized with catalytic RuO$_4$ in the presence of a very electron-rich benzene ring, which fails to be oxidized probably due to steric hindrance.[54a]

When a phenyl group is integrated inside a benzyl group protecting an alcohol, treatment with RuO$_4$ can lead to oxidation at the benzylic position. In other words, benzyl ethers are transformed into benzoates with catalytic RuO$_4$.[55] Regarding the relative sensitivity of benzyl ethers versus primary alcohols against RuO$_4$, examples are found in the literature with all possible outcomes. Thus, primary alcohols are sometimes oxidized in the presence of unreacting benzyl ethers.[56] Other times, the reverse happens and benzyl ethers are transformed into benzoates in the presence of unreacting primary alcohols.[55a] Additionally, in some cases both benzyl ethers and alcohols are oxidized.[55b]

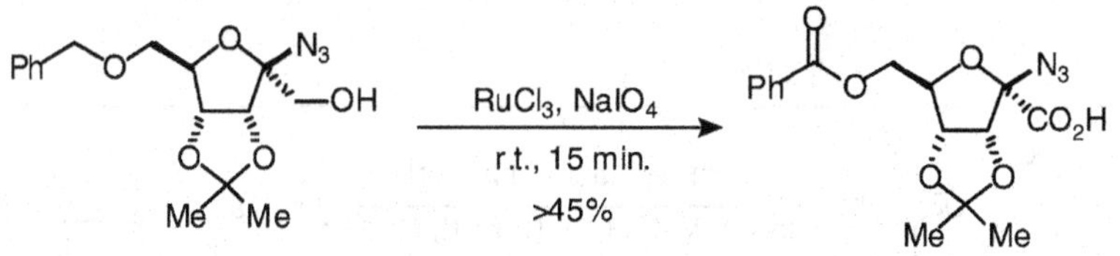

Treatment with RuO_4 leads to the simultaneous oxidation of a primary alcohol to a carboxylic acid and a benzyl ether to a benzoate.[55b]

Secondary alcohols are oxidized to ketones with RuO_4,[57] and molecules containing both secondary and primary alcohols are normally oxidized to ketoacids with RuO_4.[6, 58]

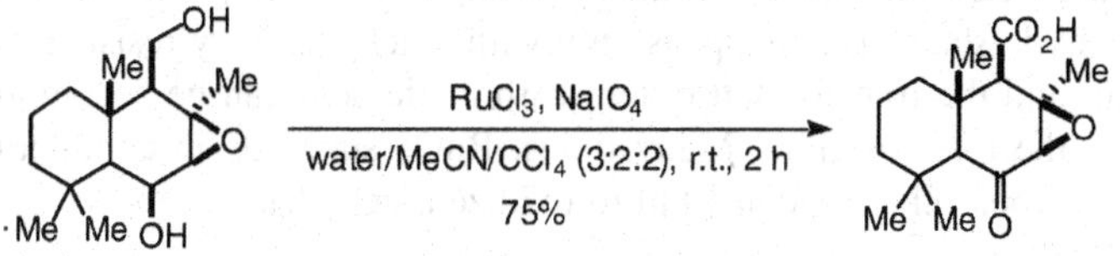

Both a primary and a secondary alcohol are oxidized with RuO_4.[58]

Interestingly, it is possible to oxidize selectively primary alcohols in the presence of secondary ones with RuO_4[8c] by adjusting carefully the reaction conditions, something that is seldom employed for synthetic advantage.

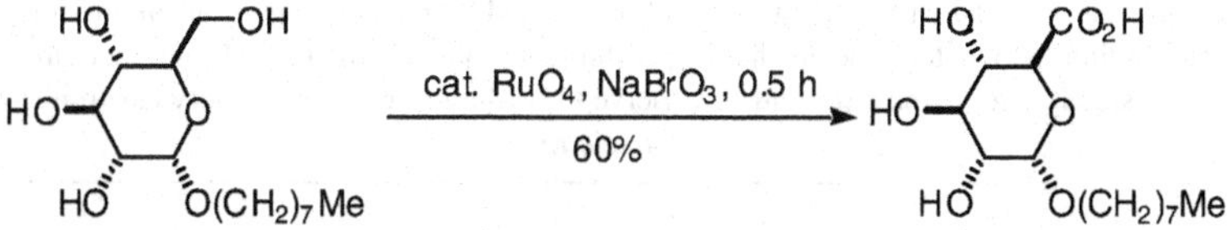

A primary alcohol is oxidized to a carboxylic acid with RuO_4 in the presence of three secondary alcohols, using RuO_4 in aqueous solution with $NaBrO_3$ as secondary oxidant. Interestingly, a much better selectivity is found using RuO_4 rather than ruthenate or perruthenate, which are weaker oxidants.[8c]

Although selective oxidations of primary alcohols are possible in the presence of other functional groups that would be oxidized by RuO$_4$ under more drastic conditions, RuO$_4$ is sometimes employed for the purposeful oxidation of several functional groups including primary alcohols.[59]

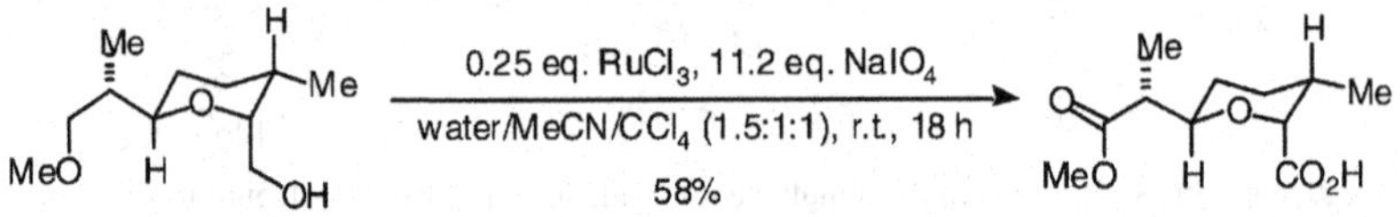

RuO$_4$ is used for the concurrent oxidation of a primary alcohol to acid and a methyl ether to ester.[59] It must be mentioned that there are very few procedures for the clean transformation of methyl ethers to methyl esters.

5.4. Side Reactions

In very rare cases, oxidation of primary alcohols with RuO$_4$ stops at the intermediate aldehyde stage. This happens only with aldehydes very resistant to further oxidation, and the transformation into carboxylic acid can be accomplished by extending the reaction time. Many times, RuO$_4$ succeeds in completing these oxidations when other oxidants fail to oxidize aldehydes.[18]

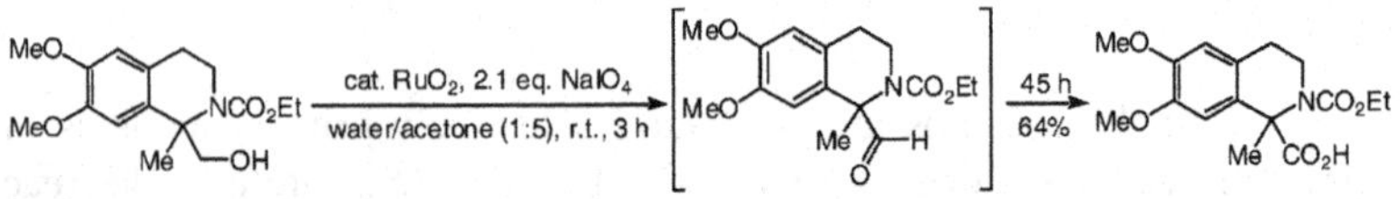

According to the authors: "The oxidation of this primary alcohol was now examined. Treatment with Jones' reagent or with a variety of other common oxidizing agents afforded the aldehyde, which was seemingly impervious to further oxidation. Eventually it was found that oxidation to the carboxylic acid could be effected by treatment with ruthenium dioxide and sodium periodate." The alcohol is transformed initially into aldehyde and then, much more slowly, into carboxylic acid.[18] Observe the failure to react of the very electron-rich aromatic ring.

A very common side reaction during the transformation of primary alcohols into carboxylic acids using many oxidizing agents consists of the epimerization at the carboxylic acid center. Compared with other oxidants, RuO$_4$ is remarkable because of its very low tendency to induce such epimerizations.[60]

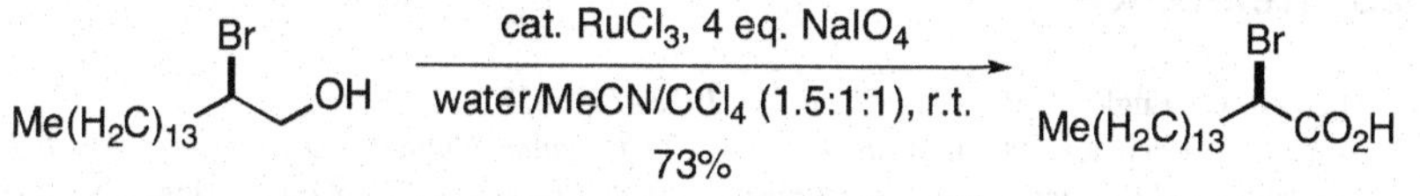

According to the authors: "Concerning the oxidation step, a major problem associated with our compound is the presence of a readily epimerizable position next to the newly formed carboxylic function. Among the methods described in the literature for the oxidation of alcohols, Sharpless oxidation was considered the most suitable, due to its well precedented compatibility with epimerizable stereogenic centers. Thus, under the standard reaction conditions for the process, [the] alcohol was oxidized to [the] bromoacid without racemization at C-2 in 84% yield." [60a]

Side products are sometimes isolated, which can be explained by intramolecular attack by nucleophiles at intermediate carbocations generated during the oxidation. [61]

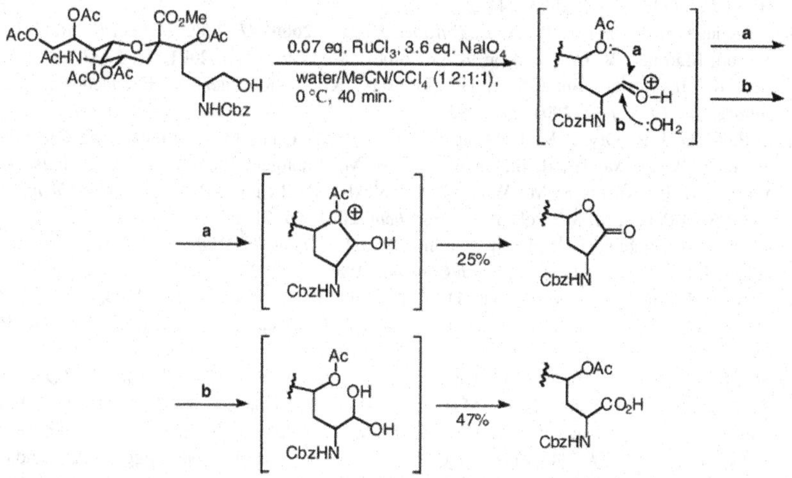

The oxidation of the primary alcohol with RuO_4 leads to a carbocation that may suffer either an intramolecular attack by a closely positioned oxygen atom (path **a**) or attack by water (path **b**). Path **a** eventually leads to the formation of a lactone, while path **b** results in the desired oxidation to carboxylic acid. The formation of the lactone can be minimized by proper tuning of the reaction conditions. Thus, according to the authors, "temperature control and the amount of the catalyst were critical in the $RuCl_3$-based oxidation step to prevent lactonization." [61]

5.5. References

1 Djerassi, C.; Engle, R. R.; *J.Am.Chem.Soc.* **1953**, *75*, 3838.

2 Tojo, G.; Fernández, M. in *Basic Reactions in Organic Synthesis. Oxidation of Alcohols to Aldehydes and Ketones: A Guide to Current Common Practice*, G. Tojo, Ed.; Springer: New York, 2006, p. 220.

3 Martin, F. S.; *J.Chem.Soc.* **1952**, 3055.

4 Pappo, R.; Becker, A.; *Bull.Res.Counc. Isr.* **1956**, *5A*, 300.

5 Berkowitz, L. M.; Rylander, P. N.; *J.Am.Chem.Soc.* **1958**, *80*, 6682.

6 Roberts, B. W.; Poonian, M. S.; Welch, S. C.; *J.Am.Chem.Soc.* **1969**, *91*, 3400.

7 Carlsen, P. H. J.; Katsuki, T.; Martin, V. S.; Sharpless, K. B.; *J.Org.Chem.* **1981**, *46*, 3936.

8 a) Lee, D. G.; Van den Engh, M.; *Can.J.Chem.* **1972**, *50*, 2000. b) Lee, D. G.; Spitzer, U. A.; Cleland, J.; Olson, M. E.; *ibid.* **1976**, *54*, 2124. c) Boelrijk, A. E. M.; Reedijk, J.; *J.Mol.Catal.* **1994**, *89*, 63.

9 a) Dolle, R. E.; Nicolaou, K. C.; *J.Am.Chem.Soc.* **1985**, *107*, 1691. b) Tachimori, Y.; Sakakibara, T.; Sudoh, R.; *Carbohydr.Res.* **1981**, *95*, 299. c) Baldwin, J. E.; Killin, S. J.; Adlington, R. M.; Spiegel, U.; *Tetrahedron* **1988**, *44*, 2633. d) Herdeis, C.; Hubmann, H. P.; *Tetrahedron: Asymmetry* **1992**, *3*, 1213. e) Burgess, K.; Liu, L. T.; Pal, B.; *J.Org.Chem.* **1993**, *58*, 4158. f) Hart, B. P.; Verma, S. K.; Rapoport, H.; *J.Org.Chem.* **2003**, *68*, 187. g) Park, J.-il; Tian, G. R.; Kim, D. H.; *J.Org.Chem.* **2001**, *66*, 3696. h) Lucas, B. S.; Luther, L. M.; Burke, S. D.; *Org.Lett.* **2004**, *6*, 2965.

10 Kriek, N. M. A. J.; van der Hout, E.; Kelly, P.; van Meijgaarden, K. E.; Geluk, A.; Ottenhoff, T. H. M.; van der Marel, G. A.; Overhand, M.; van Boom, J. H.; Valentijn, A. R. P. M.; Overkleeft, H. S.; *Eur.J.Org.Chem.* **2003**, 2418.

11 a) Godage, H. Y.; Fairbanks, A. J.; *Tetrahedron Lett.* **2000**, *41*, 7589. b) Lee, K.; Ravi, G.; Ji, X.-duo; Márquez, V. E.; Jacobson, K. A.; *Bioorg.Med.Chem.Lett.* **2001**, *11*, 1333. c) Ravi, G.; Lee, K.; Ji, X.-duo; Sung Kim, H.; Soltysiak, K. A.; Márquez, V. E.; Jacobson, K. A.; *Bioorg.Med.Chem.Lett.* **2001**, *11*, 2295.

12 a) Pinto, D. J. P.; Orwat, M. J.; Wang, S.; Fevig, J. M.; Quan, M. L.; Amparo, E.; Cacciola, J.; Rossi, K. A.; Alexander, R. S.; Smallwood, A. M.; Luettgen, J. M.; Liang, L.; Aungst, B. J.; Wright, M. R.; Knabb, R. M.; Wong, P. C.; Wexler, R. R.; Lam, P. Y. S.; *J.Med.Chem.* **2001**, *44*, 566. b) Palian, M. M.; Polt, R.; *J.Org.Chem.* **2001**, *66*, 7178.

13 Khan, F. A.; Prabhudas, B.; Dash, J.; Sahu, N.; *J.Am.Chem.Soc.* **2000**, *122*, 9558.

14 Genet, J. P.; Pons, D.; Jugé, S.; *Synth.Commun.* **1989**, *19*, 1721.

15 Cornely, J.; Su Ham, L. M.; Meade, D. E.; Dragojlovic, V.; *Green Chem.* **2003**, *5*, 34.

16 Prashad, M.; Lu, Y.; Kim, H.-Y.; Hu, B.; Repic, O.; Blacklock, T. J.; *Synth.Commun.* **1999**, *29*, 2937.

17 See, for example: a) Eaton, P. E.; Cooper, G. F.; Johnson, R. C.; Mueller, R. H.; *J.Org.Chem.* **1972**, *37*, 1947. b) Eaton, P. E.; Mueller, R. H.; *J.Am.Chem.Soc.* **1972**, *94*, 1014. c) Flores-Parra, A.; Khuong-Huu, F.; *Tetrahedron* **1986**, *42*, 5925. d) Ziegler, F. E.; Cain, W. T.; *J.Org.Chem.* **1989**, *54*, 3347. e) Bravo, P.; Cavicchio, G.; Crucianelli, M.; Poggiali, A.; Zanda, M.; *Tetrahedron: Asymmetry* **1997**, *8*, 2811.

18 Czarnocki, Z.; Suh, D.; Maclean, D. B.; Hultin, P. G.; Szarek, W. A.; *Can.J.Chem.* **1992**, *70*, 1555.

19 Garner, P.; Park, J. M.; *J.Org.Chem.* **1990**, *55*, 3772.

20 a) Mori, K.; Ebata, T.; *Tetrahedron* **1986**, *42*, 4413. b) Otaka, A.; Miyoshi, K.; Burke Jr., T. R.; Roller, P. P.; Kubota, H.; Tamamura, H.; Fujii, N.; *Tetrahedron Lett.* **1995**, *36*, 927.

21 a) Bonini, C.; Di Fabio, R.; *Tetrahedron Lett.* **1988**, *29*, 815. b) Pastó, M.; Moyano, A.; Pericàs, M. A.; Riera, A.; *Tetrahedron: Asymmetry* **1996**, *7*, 243. c) Pastó, M.; Castejón, P.; Moyano, A.; Pericàs, M. A.; Riera, A.; *J.Org.Chem.* **1996**, *61*, 6033. d) Plake, H. R.; Sundberg, T. B.; Woodward, A. R.; Martin, S. F.; *Tetrahedron Lett.* **2003**, *44*, 1571. e) Jain, R. P.; Vederas, J. C.; *Org.Lett.* **2003**, *5*, 4669.

22 Denis, J.-N.; Greene, A. E.; Aarão Serra, A.; Luche, M.-J.; *J.Org.Chem.* **1986**, *51*, 46.

23 Šarek, J.; Klinot, J.; Džubák, P.; Klinotová, E.; Nosková, V.; Křeček, V.; Kořínková, G.; Thomson, J. O.; Janošt'áková, A.; Wang, S.; Parsons, S.; Fischer, P. M.; Zhelev, N. Z.; Hajdúch, M.; *J.Med.Chem.* **2003,** *46,* 5402.

24 Parikh, V. M.; Jones, J. K. N.; *Can. J. Chem.* **1965,** *43,* 3452.

25 Falorni, M.; Porcheddu, A.; Giacomelli, G.; *Tetrahedron: Asymmetry* **1997,** *8,* 1633.

26 Morea, G.; Sabbatini, L.; Zambonin, P. G.; Tangari, N.; Tortorella, V.; *J.Chem.Soc. Faraday Trans. I* **1989,** *85,* 3861.

27 Franchini, C.; Carocci, A.; Catalano, A.; Cavalluzzi, M. M.; Corbo, F.; Lentini, G.; Scilimati, A.; Tortorella, P.; Conte Camerino, D.; De Luca, A.; *J.Med.Chem.* **2003,** *46,* 5238.

28 Chong, J. M.; Sharpless, K. B.; *J.Org.Chem.* **1985,** *50,* 1560.

29 Stock, L. M.; Tse, K.-tuen; *Fuel* **1983,** *62,* 974.

30 a) Witiak, D. T.; Patch, R. J.; Enna, S. J.; Fung, Y. K.; *J.Med.Chem.* **1986,** *29,* 1. b) Thaisrivongs, S.; Pals, D. T.; Kroll, L. T.; Turner, S. R.; Han, F.-S.; *J.Med.Chem.* **1987,** *30,* 976. c) Núñez, M. T.; Martin, V. S.; *J.Org.Chem.* **1990,** *55,* 1928. d) Mawson, S. D.; Weavers, R. T.; *Tetrahedron* **1995,** *51,* 11257. e) Xue, C.-B.; Voss, M. E.; Nelson, D. J.; Duan, J. J.-W.; Cherney, R. J.; Jacobson, I. C.; He, X.; Roderick, J.; Chen, L.; Corbett, R. L.; Wang, L.; Meyer, D. T.; Kennedy, K.; DeGrado, W. F.; Hardman, K. D.; Teleha, C. A.; Jaffee, B. D.; Liu, R.-Q.; Copeland, R. A.; Covington, M. B.; Christ, D. D.; Trzaskos, J. M.; Newton, R. C.; Magolda, R. L.; Wexler, R. R.; Decicco, C. P.; *J.Med.Chem.* **2001,** *44,* 2636. f) Mühlman, A.; Classon, B.; Hallberg, A.; Samuelsson, B.; *J.Med.Chem.* **2001,** *44,* 3402. g) Venkatraman, S.; Njoroge, F. G.; Girijavallabhan, V.; McPhail, A. T.; *J.Med.Chem.* **2002,** *45,* 2686.

31 Barak, G.; Dakka, J.; Sasson, Y.; *J.Org.Chem.* **1988,** *53,* 3553.

32 Lee, D. G.; Hall, D. T.; Cleland, J. H.; *Can.J.Chem.* **1972,** *50,* 3741.

33 Schröder, M.; Griffith, W. P.; *J.Chem.Soc. Chem.Commun.* **1979,** 58.

34 Varma, R. S.; Hogan, M. E.; *Tetrahedron Lett.* **1992,** *33,* 7719.

35 a) Filippov, D.; Meeuwenoord, N. J.; van der Marel, G. A.; Efimov, V. A.; Kuyl-Yeheskiely, E.; van Boom, J. H.; *Synlett* **1996,** 769. b) Fonquerna, S.; Rios, R.; Moyano, A.; Pericàs, M. A.; Riera, A.; *Eur.J.Org.Chem.* **1999,** 3459. c) Soo Kim, K.; Joo Lee, Y.; Hwan Kim, J.; Kyung Sung, D.; *Chem.Commun.* **2002,** 1116.

36 Tojo, G.; Fernández, M. in *Basic Reactions in Organic Synthesis, Oxidation of Alcohols to Aldehydes and Ketones: A Guide to Current Common Practice,* G. Tojo, Ed.; Springer: New York, 2006, section 4.3.

37 Benningshof, J. C. J.; Ijsselstijn, M.; Wallner, S. R.; Koster, A. L.; Blaauw, R. H.; van Ginkel, A. E.; Brière, J.-F.; van Maarseveen, J. H.; Rutjes, F. P. J. T.; Hiemstra, H.; *J.Chem.Soc. Perkin Trans. I* **2002,** 1701.

38 Greenfield, A. A.; Butera, J. A.; Caufield, C. E.; *Tetrahedron Lett.* **2003,** *44,* 2729.

39 a) Houri, A. F.; Xu, Z.; Cogan, D. A.; Hoveyda, A. H.; *J.Am.Chem.Soc.* **1995,** *117,* 2943. b) Hu, T.; Panek, J. S.; *J.Am.Chem.Soc.* **2002,** *124,* 11368.

40 Boger, D. L.; Ledeboer, M. W.; Kume, M.; *J.Am.Chem.Soc.* **1999,** *121,* 1098.

41 Still, W. C.; Ohmizu, H.; *J.Org.Chem.* **1981,** *46,* 5242.

42 Guibourdenche, C.; Roumestant, M. L.; Viallefont, Ph.; *Tetrahedron: Asymmetry* **1993,** *4,* 2041.

43 a) Albarella, L.; Piccialli, V.; Smaldone, D.; Sica, D.; *J.Chem.Res. (S)* **1996,** *9,* 400. b) Plietker, B.; Niggemann, M.; *Org.Lett.* **2003,** *5,* 3353. c) Wolfe, S.; Hasan, S. K.; Campbell, J. R.; *J.Chem.Soc. Chem.Commun.* **1970,** 1420. d) Almansa, C.; Carceller, E.; Moyano, A.; Serratosa, F.; *Tetrahedron* **1986,** *42,* 3637. e) Griffith, W. P.; Kwong, E.; *Synth.Commun.* **2003,** *33,* 2945.

44 Caputo, J. A.; Fuchs, R.; *Tetrahedron Lett.* **1967,** *8,* 4729.

45 a) Denmark, S. E.; Cramer, C. J.; Sternberg, J. A.; *Tetrahedron Lett.* **1970,** 4003. b) Denmark, S. E.; Cramer, C. J.; Sternberg, J. A.; *Helv.Chim.Acta* **1986,** *69,* 1971.

46 Wakamatsu, K.; Kigoshi, H.; Niiyama, K.; Niwa, H.; Yamada, K.; *Tetrahedron* **1986,** *42,* 5551.

47 Sicinski, R. R.; DeLuca, H. F.; *Biorg.Med.Chem.Lett.* **1995,** *5,* 159.

48 Chang, M.-Y.; Chen, C.-Y.; Tasi, M.-R.; Tseng, T.-W.; Chang, N.-C.; *Synthesis* **2004,** 840.

49 a) Chakraborty, T. K.; Ghosh, A.; *Tetrahedron Lett.* **2002**, *43*, 9691. b) Hanselmann, R.; Zhou, J.;
    Ma, P.; Confalone, P. N.; *J.Org.Chem.* **2003**, *68*, 8739. c) Charrier, J.-D.; Hitchcock, P. B.;
    Young, D. W.; *Org.Biomol.Chem.* **2004**, *2*, 1310.

50 a) Godier-Marc, E.; Aitken, D. J.; Husson, H.-P.; *Tetrahedron Lett.* **1997**, *38*, 4065.
    b) Brickmann, K.; Yuan, Z.Q.; Sethson, I.; Somfai, P.; Kihlberg, J.; *Chem.Eur.J.* **1999**, *5*, 2241.
    c) Okamoto, N.; Hara, O.; Makino, K.; Hamada, Y.; *Tetrahedron: Asymmetry* **2001**, *12*, 1353.
    d) Anwar, M.; Bailey, J. H.; Dickinson, L. C.; Edwards, H. J.; Goswami, R.; Moloney, M. G.
    Org.Biomol.Chem. **2003**, *1*, 2364.

51 Ghosh, S. K.; Satapathi, T. K.; Subba Rao, P. S. V.; Sarkar, T.; *Synth.Commun.* **1988**, *18*, 1883.

52 a) Ermert, P.; Meyer, J.; Stucki, C.; Schneebeli, J.; Obrecht, J.-P.; *Tetrahedron Lett.* **1988**, *29*, 1265.
    b) Cook, G. R.; Shanker, P. S.; *Tetrahedron Lett.* **1998**, *39*, 3405. c) Whisler, M. C.; Beak, P.;
    J.Org.Chem. **2003**, *68*, 1207. d) Nazih, A.; Schneider, M.-R.; Mann, A.; *Synlett* **1998**, 1337.

53 Song, Z. J.; Zhao, M.; Desmond, R.; Devine, P.; Tschaen, D. M.; Tillyer, R.; Frey, L.; Heid, R.;
    Xu, F.; Foster, B.; Li, J.; Reamer, R.; Volante, R.; Grabowski, E. J. J.; Dolling, U. H.; Reider,
    P. J.; Okada, S.; Kato, Y.; Mano, E.; *J.Org.Chem.* **1999**, *64*, 9658.

54 a) Moutel, S.; Prandi, J.; *J.Chem.Soc. Perkin Trans. I* **2001**, 305. b) Crilley, M. M. L.; Edmunds,
    A. J. F.; Eistetter, K.; Golding, B. T.; *Tetrahedron Lett.* **1989**, *30*, 885.

55 a) Takeda, R.; Zask, A.; Nakanishi, K.; Park, M. H.; *J.Am.Chem.Soc.* **1987**, *109*, 914. b) Gasch, C.;
    Salameh, B. A. B.; Pradera, M. A.; Fuentes, J.; *Tetrahedron Lett.* **2001**, *42*, 8615. c) Schuda,
    P. F.; Cichowicz, M. B.; Heimann, M. R.; *Tetrahedron Lett.* **1983**, *24*, 3829.

56 a) Tanner, D.; Somfai, P.; *Tetrahedron Lett.* **1988**, *29*, 2373. b) Koulocheri, S. D.; Magiatis, P.;
    Skaltsounis, A.-L.; Haroutounian, S. A.; *Tetrahedron* **2002**, *58*, 6665. c) Ghosh, A. K.; Lei, H.;
    J.Org.Chem. **2002**, *67*, 8783.

57 Tojo, G.; Fernández, M. in *Basic Reactions in Organic Synthesis, Oxidation of Alcohols to
    Aldehydes and Ketones: A Guide to Current Common Practice*, G. Tojo, Ed.; Springer: New
    York, 2006, section 4.2..

58 Lagnel, B. M. F.; de Groot, A.; Morin, C.; *Tetrahedron Lett.* **2001**, *42*, 7225.

59 Whitney, R. A.; *Can.J.Chem.* **1986**, *64*, 803.

60 a) Hernanz, D.; Camps, F.; Guerrero, A.; Delgado, A.; *Tetrahedron: Asymmetry* **1995**, *6*, 2291.
    b) Yokomatsu, T.; Sato, M.; Shibuya, S.; *Tetrahedron: Asymmetry* **1996**, *7*, 2743.

61 Wang, Q.; Linhardt, R. J.; *J.Org.Chem.* **2003**, *68*, 2668.

TEMPO-Mediated Oxidations

6.1. Introduction

In 1965, Golubev, Rozantsev, and Neiman reported[1] that treatment of oxoam-monium salt **11** with excess of ethanol led to the formation of acetaldehyde.

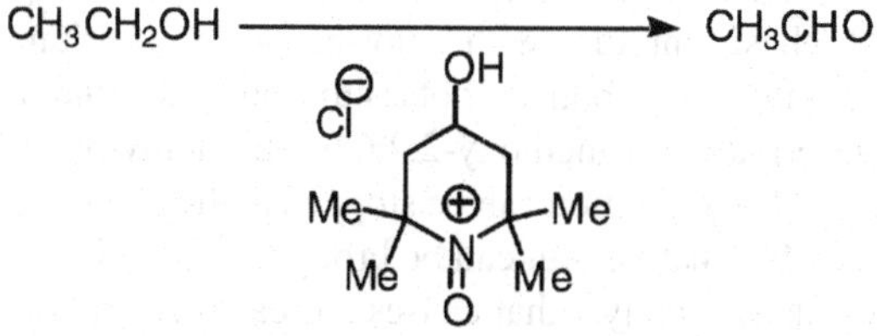

CH$_3$CH$_2$OH ⟶ CH$_3$CHO

11

In 1975, Cella et al. demonstrated[2] that alcohols can be oxidized to carboxylic acids by treatment with *m*-chloroperbenzoic acid in the presence of a catalytic amount of 2,2,6,6-tetramethylpiperidine (**12**).

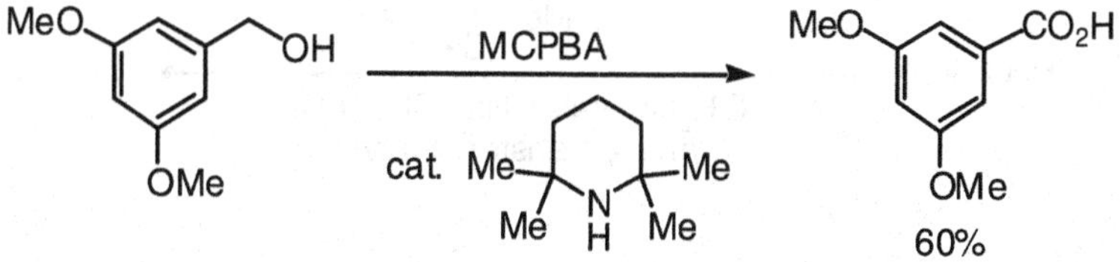

MCPBA

cat. **12**

60%

Apparently, MCPBA oxidizes the amine **12**, resulting in a catalytic quantity of the stable radical 2,2,6,6-tetramethylpiperidine-1-oxyl (**13**), normally called TEMPO, that is further oxidized to the oxoammonium cation **14** that operates as the primary oxidant.

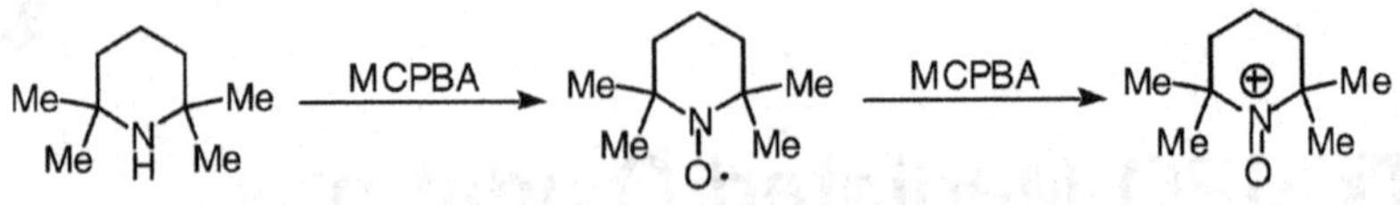

TEMPO (13)   14

Cella made a very important seminal contribution to the TEMPO-mediated obtention of carboxylic acids by showing that oxoammonium salts can be employed catalytically in the transformation of primary alcohols into carboxylic acids. On the other hand, Cella's procedure involves the use of a peracid as secondary oxidant, which is a strong oxidant that interferes with many functional groups.

In 1987, Anelli *et al.* published[3] a landmark paper in which they showed that primary alcohols can be oxidized either to aldehydes or to carboxylic acids in a highly efficient and convenient manner, by treating the alcohol in a CH_2Cl_2–water biphasic mixture with chlorine bleach (sodium hypochlorite) in the presence of sodium bicarbonate, potassium bromide, and a catalytic amount of the TEMPO derivative 4-methoxy-2,2,6,6-tetramethylpiperidine-1-oxyl (**15**) (4-MeO-TEMPO). The oxidation can be stopped at the aldehyde stage by running it for a short time. Alternatively, it can be brought to the carboxylic acid stage by adding a phase-transfer catalyst that causes a great acceleration of the oxidation.

Anelli`s oxidation of primary alcohols to carboxylic acids

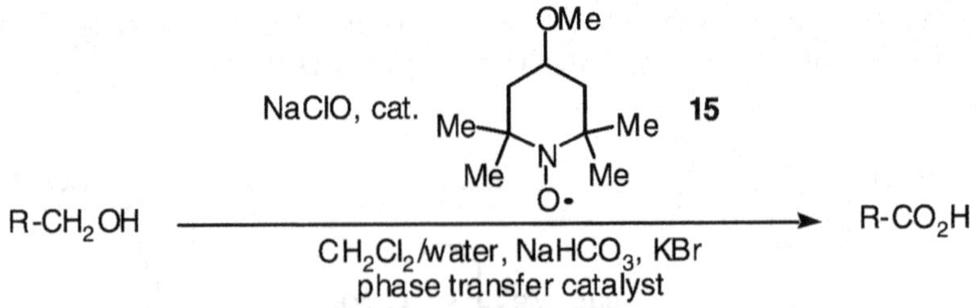

One important limitation of Anelli's procedure is the need to utilize sodium hypochlorite as stoichiometric oxidant, a compound that has a great tendency to produce chlorinations in some sensitive substrates. This tendency to chlorination can be mitigated by employing Zhao's modification of Anelli's procedure, which was published[4] in 1999. In this modification, the secondary oxidant—sodium hypochlorite—is used in catalytic amounts rather than in excess, the reagent being regenerated by addition of stoichiometric sodium chlorite, a compound that lacks the strong chlorinating tendency of sodium hypochlorite.

Zhao`s modification of Anelli`s oxidation

$$R\text{-}CH_2OH \xrightarrow[\text{MeCN/sodium phophate buffer}]{NaClO_2, \text{ cat. } NaClO, \text{ cat. } TEMPO} R\text{-}CO_2H$$

In 1999, Epp and Widlanski described[5] the oxidation of alcohols to carboxylic acids using catalytic TEMPO, with bis(acetoxy)iodobenzene—PhI(OAc)$_2$, commonly referred as BAIB—as secondary oxidant in an acetonitrile–aqueous buffer mixture. This procedure for the oxidation of primary alcohols possesses the distinctive advantage of producing the rather benign iodobenzene and acetic acid as side compounds. Furthermore, in contrast to other oxidation procedures, it is possible to perform the oxidation of Epp and Widlanski in the absence of metallic salts.

Procedure of Epp and Widlanski for oxidation of primary alcohols to carboxylic acids

$$R\text{-}CH_2OH \xrightarrow[\text{MeCN/aqueous buffer}]{\text{cat. TEMPO, PhI(OAc)}_2} R\text{-}CO_2H$$

Mechanism

The available experimental data are consistent with a mechanism,[4, 6] as shown below, in which the secondary oxidant transforms TEMPO, or a related stable radical, in an oxoammonium salt that operates as the primary oxidant, transforming the alcohol into the corresponding aldehyde. This results in the formation of a hydroxylamine that is oxidized to a TEMPO radical, thus completing the catalytic cycle.

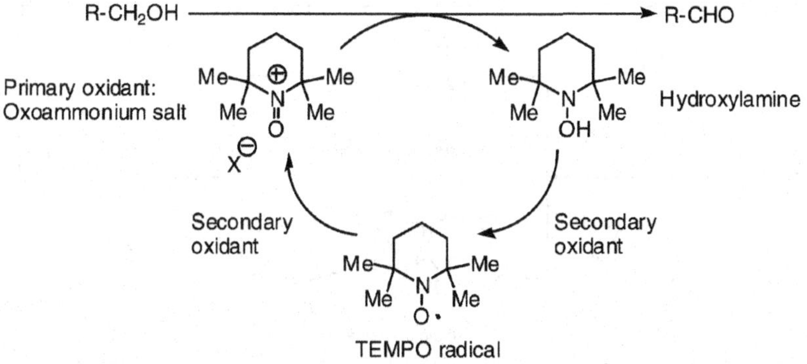

The catalytic cycle can in fact be more complex, because TEMPO radicals can dispro-portionate into oxoammonium salts and hydroxylamines under acidic catalysis.[6a]

The aldehyde, in the presence of water, equilibrates with the corresponding hydrate that can be oxidized via a similar mechanism to the corresponding acid, as shown below.

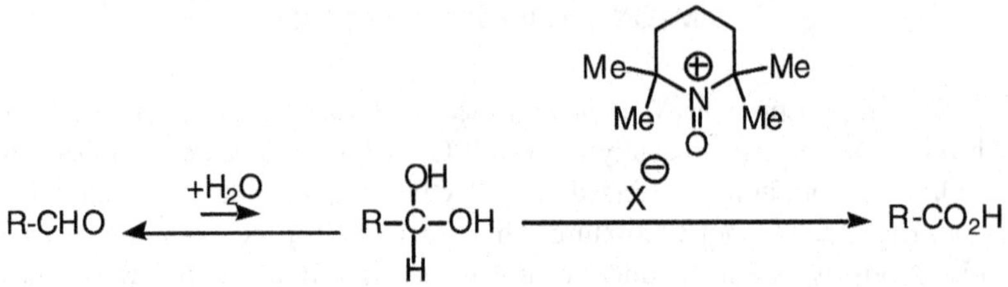

The oxoammonium salt operates as primary
oxidant in a catalytic cycle as above

Interestingly, TEMPO inhibits the oxidation of aldehydes to carboxylic acids when this oxidation proceeds via a radical mechanism. That is why Anelli's oxidation can be carried out under air and be easily stopped at the aldehyde stage with no competing overoxidation due to the presence of gaseous oxygen.[7]

While in all TEMPO-mediated oxidations of primary alcohols to carboxylic acids, oxoammonium salts are the primary oxidants for the transformation of alcohols into aldehydes, the subsequent oxidation of aldehydes to carboxylic acids may sometimes be effected by the oxidant present in excess rather than by oxoammonium salts. In such cases, the secondary oxidant for the transformation of alcohols into aldehydes is the primary oxidant for the oxidation of aldehydes to carboxylic acids.

The oxidation of primary alcohols with oxoammonium salts can work either via a compact five-membered transition state under basic conditions or via a linear transition state under acidic conditions, as shown below. Under basic conditions the oxidation is quicker and possesses a greater selectivity for primary alcohols versus secondary ones.

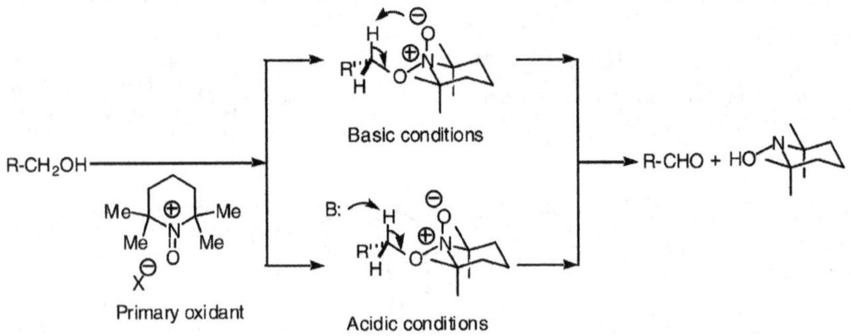

The five-membered transition state under basic conditions is more compact, leading to a quicker reaction rate and greater selectivity for oxidation of primary alcohols relative to secondary ones.

Stoichiometric Oxidants

The most common stoichiometric oxidants in TEMPO-mediated transformations of primary alcohols into carboxylic acids are sodium hypochlorite (NaOCl)—Anelli's oxidation—,[3] sodium chlorite (NaClO$_2$)—Zhao's modification of Anelli's oxidation—,[4] and PhI(OAc)$_2$—oxidation of Epp and Widlanski.[5] Other stoichiometric oxidants less commonly used include MCPBA,[2] Ca(ClO)$_2$ (swimming pool bleach),[8] t-BuOCl,[9] CuCl – O$_2$,[10] NaBrO$_2$,[11] Cl$_2$,[12] Br$_2$[12], and trichloroisocyanuric acid.[13] It is possible to perform the oxidation under electrochemical conditions in the presence of catalytic TEMPO.[14]

Amino 1-Oxyl Radicals

TEMPO and analogue compounds can be prepared starting with a simple condensation of ammonia with acetone,[15] and their cost is quite affordable. They are stable radicals because they are flanked by two quaternary carbons that provide a bulky environment. For these reasons, TEMPO and analogue compounds, that is, 2,2,6,6-tetramethylpiperidine-1-oxyl radicals, are almost exclusively employed in oxidations mediated by amino 1-oxyl radicals.

> TEMPO is a volatile compound soluble in organic solvents and water. It can be recovered by extraction with Et$_2$O[14] or by azeotropic distillation with water.[16]

Although some differences in the chemical behavior of different TEMPO derivatives were noticed,[17] the selection of a particular derivative is normally dictated by price and convenience. As far as the authors are aware, no profound study of the efficacy of different TEMPO derivatives in the oxidation of alcohols has been carried out.

> TEMPO entrapped within a silica matrix has been employed as a recyclable catalyst in the selective oxidation of primary alcohols using NaOCl as stoichiometric oxidant.[18]

6.2. Anelli's Oxidation

In 1987, Anelli *et al.* made a key contribution to TEMPO-mediated oxidations by showing that the very cheap reagent chlorine bleach (aqueous NaOCl) can function very effectively as a stoichiometric oxidant for alcohols in the presence of traces of 4-MeO-TEMPO.[3] They established a protocol involving a reaction run at 0 °C in a biphasic CH$_2$Cl$_2$–water mixture in the presence of excess of NaOCl, NaHCO$_3$, KBr, and catalytic 4-MeO-TEMPO for oxidation to aldehydes. Under these conditions the oxidation to acids is quite slow. If the acid is desired, it is advisable to add a phase-transfer catalyst to speed up the oxidation.

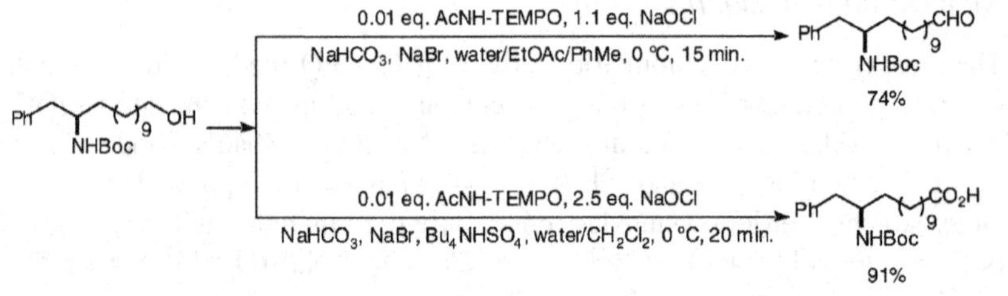

The oxidation of a primary alcohol using Anelli's procedure *without* the addition of a phase-transfer catalyst allows the preparation of 74% of the corresponding aldehyde. The addition of 1 mol% of tetrabutylammonium bisulfate as phase-transfer catalyst produces a great acceleration of the oxidation that allows the isolation of the corresponding acid in 91%.[19]

The following experimental data are relevant regarding Anelli's oxidation:

- Primary alcohols are transformed into the corresponding aldehydes—with no need to add a phase-transfer catalyst—normally in only about 3 minutes. The oxidation of benzyl alcohols possessing electron donating groups in the aromatic ring can be much slower, a fact that can be explained by the presence of a negative charge in the transition state for the oxidation involving oxoammonium salts under basic conditions.

- The addition of some KBr produces a substantial acceleration of the oxidation, because of the generation of HOBr.[20] This reagent is formed from HOCl and KBr, and is apparently a much better oxidant for the regeneration of oxoammonium salts than HOCl.

- The reaction is rather slow at the pH of commercial bleach (ca. 12.7), being much quicker at a pH of ca. 8.6, generated by the addition of $NaHCO_3$ as buffer. This fact can be explained assuming that at a very high pH the regeneration of the oxoammonium salt, rather than the oxidation of the alcohol by the oxoammonium salt via a five-membered transition state, becomes rate-determining. At a very basic pH the concentration of HOBr, which is the oxidant regenerating the oxoammonium salt, becomes very low relative to the concentration of the hypobromite anion (BrO^-).

- The reaction can fail in substrates possessing a high hydrophilicity. Apparently, the oxidation takes place in the organic phase, where such substrates are present in a very low concentration.

- The oxidation can be substantially accelerated by the addition of quaternary ammonium salts as a phase-transfer catalyst. Thus, while in the absence of phase-transfer catalyst the reaction is easily stopped at the aldehyde stage, the addition of catalytic amounts of a quaternary ammonium salt normally allows the oxidation to carboxylic acid to be completed in 5 minutes at 0 °C. As in the oxidation to aldehydes, electronic effects can be very important, and oxidations yielding benzoic acids possessing electron donating groups can be much slower.

• Somehow unexpectedly, the reaction speed decreases by increasing the temperature, a fact due to the decomposition of oxoammonium salts, which are very stable at 0 °C, but decompose very quickly in the presence of water at 25 °C.

• The transformation of aldehydes into carboxylic acids is apparently mediated by oxoammonium salts, rather than by some other oxidant in excess; for, in the absence of TEMPO radicals, this reaction is rather slow.

The above facts are illustrated in the following scheme:

Oxidation of primary alcohols into carboxylic acids by Anelli`s procedure

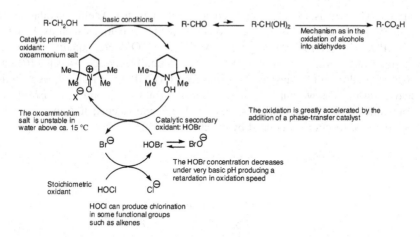

Solvent

Anelli's oxidation is most often carried out in a biphasic system consisting of CH_2Cl_2 and water. As the oxidation takes place in the organic phase, this biphasic system fails in water-soluble substrates. That is why when Anelli's oxidation is applied to sugars, water is normally employed as the sole solvent.[21] Furthermore, acetone/water is a solvent combination quite often utilized,[22] while

MeCN/water[23] and EtOAc/water[24] are less used. When water or an aqueous mixture is employed as solvent, the reaction may fail in lipophilic substrates because of lack of solubility.[25]

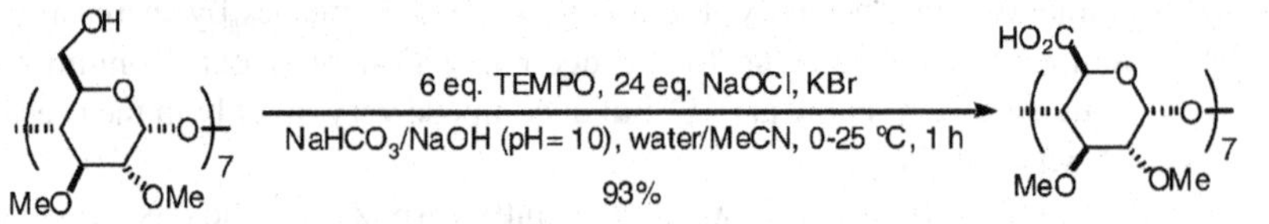

6 eq. TEMPO, 24 eq. NaOCl, KBr

NaHCO$_3$/NaOH (pH= 10), water/MeCN, 0-25 °C, 1 h

93%

The primary alcohols, in a cyclodextrin possessing the secondary alcohols protected as methyl ethers, are oxidized in a monophasic MeCN–water mixture providing 93% of the corresponding heptacarboxylate. Under these conditions there is no oxidation in substrates where the secondary alcohols are protected as benzyl or allyl ethers, due to their much greater lipophilicity.[25]

Catalyst

Simple TEMPO is usually employed as catalyst in the Anelli's oxidation of primary alcohols to acids, although other TEMPO derivatives such as 4-MeO-TEMPO,[3, 26] 4-HO-TEMPO,[27] or 4-AcNH-TEMPO[19] are equally effective. A trace quantity of 1 mol% of catalyst is normally enough for an efficient oxidation, although because of the low price of TEMPO and its very easy elimination during the workup, the use of ca. 4–10 mol% is common. TEMPO is sometimes added in amounts as high as 1–1.5 equivalents; this may help to prevent side reactions produced by excess of HOCl.[22a,b, 28]

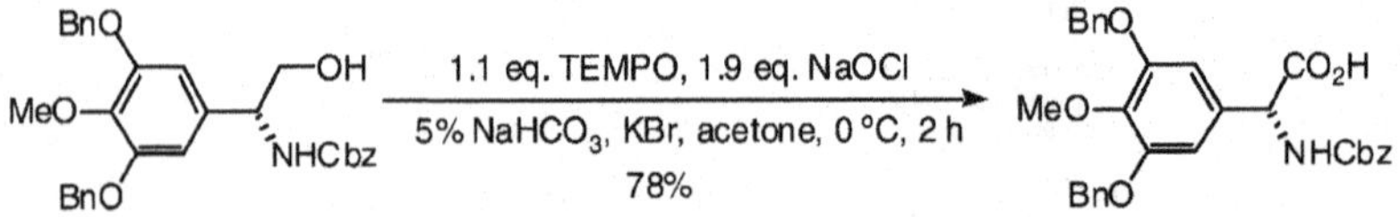

1.1 eq. TEMPO, 1.9 eq. NaOCl

5% NaHCO$_3$, KBr, acetone, 0 °C, 2 h

78%

According to the authors: "Direct oxidation of the primary alcohol to the desired sensitive carboxylic acid was accomplished best using N-oxoammonium salts in combination with NaOCl in a buffered solution (2 equivalents of 4–6% NaOCl, 1.1 equivalents of TEMPO, 0.1 equivalents of KBr, acetone-5% aqueous NaHCO$_3$, 0 °C, 2 h, 78%). In the optimization of this reaction it was found that 1.1 equivalents of TEMPO were necessary to obtain the desired oxidation product. If a catalytic amount (ca. 0.1 equivalents) of TEMPO was employed or Ca(OCl)$_2$ was substituted for NaOCl, the chlorinated aromatic derivative was isolated as the major product. Presumably the TEMPO scavenges any chlorine which is liberated during the reaction." Observe that no phase-transfer catalyst is needed in this reaction performed in a single acetone–water phase.[22a]

Phase-Transfer Catalyst

Although no phase-transfer catalyst is needed in oxidations carried out in a single phase—such as in water, MeCN/water, or acetone/water— an ammonium salt such as n-Bu$_4$NCl,[29] n-Bu$_4$NBr,[30] n-Bu$_4$NHSO$_4$,[19] or Aliquat® 336 (tricaprylylmethylammonium chloride)[31] is normally added in oxidations performed in a biphasic CH$_2$Cl$_2$/water mixture.

pH

Commercial chlorine bleach is prepared by reacting chlorine with an aqueous NaOH solution, and contains ca. 3–6% of NaOCl. Excess of NaOH is employed to stabilize the NaOCl, which otherwise would disproportionate into NaCl and NaClO$_3$. This results in bleach possessing a pH between 11 and 13, which is too basic for a normal Anelli's oxidation. The addition of NaHCO$_3$ allows lowering the pH to a value of ca. 8.6–10, which is normally ideal for Anelli's oxidations. Some NaOH is sometimes added additionally to fine tune the pH to a value of ca. 10.[25, 32]

    Figure 3 shows the influence of pH on the rate of oxidation of methyl α-D-glucopyranoside. There is a very sharp increase in speed from pH 8 to pH 10, while a higher pH produces no further acceleration of reaction speed.

    The generation of carboxylic acid during the oxidation may cause a lowering of pH that produces a decrease of oxidation speed and of selectivity

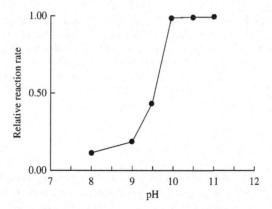

Figure 3. Influence of pH on the rate of oxidation of methyl α-D-glucopyranoside using 0.007 eq. of TEMPO, 2.2 eq. of a 15% NaOCl solution, and 0.4 eq. of NaBr at 2 °C, adding 4 M HCl to adjust the pH.[6b] Reprinted from Carbohydrate Research, Vol.269, Nooy, A.E.J.; Besemer, A.C.; van Bekkum, H., "Highly selective nitroxyl radical-mediated oxidation of primary alcohol groups in water-soluble glucans," pages 89–98, ©1995, with permission from Elsevier.

for oxidation of primary alcohols versus secondary ones. This may be avoided by continuous adjustment of the pH by addition of NaOH.[21b, 33]

Some functional groups, such as esters, may be sensitive to the mildly basic pH normally used in Anelli's oxidations, and in some cases it may be necessary to adjust the precise reaction conditions to avoid interferences.[29a]

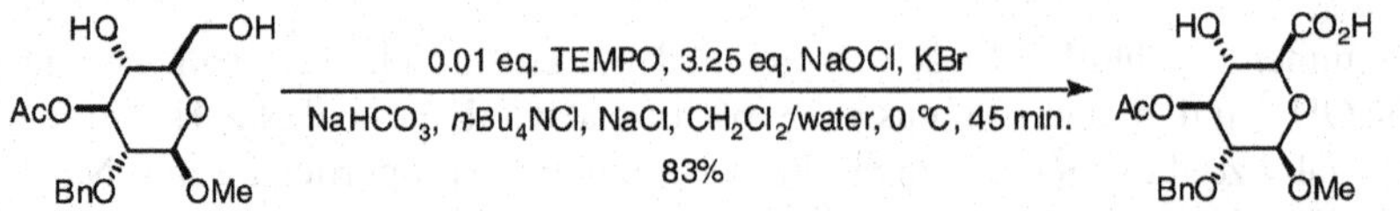
HO
OH
AcO
O
BnO
OMe
0.01 eq. TEMPO, 3.25 eq. NaOCl, KBr
NaHCO3, n-Bu4NCl, NaCl, CH2Cl2/water, 0 °C, 45 min.
83%
HO
CO2H
AcO
O
BnO
OMe

The reaction time must be kept to a minimum to avoid the hydrolysis of the acetate under the basic reaction conditions.[29a]

In some cases, Anelli's oxidation must be carried out under acidic conditions to obtain an optimum yield.[27]

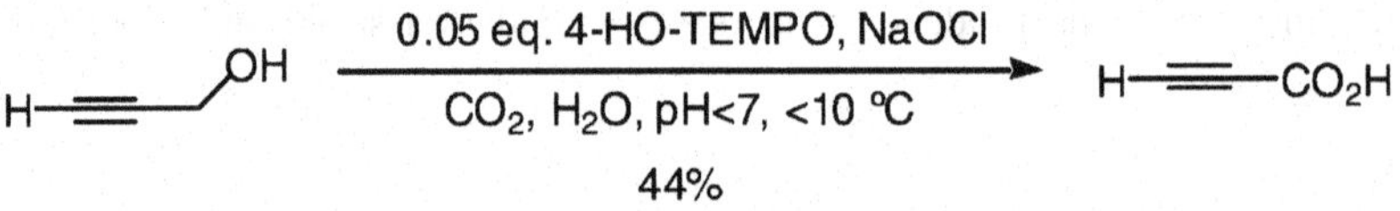
H
OH
0.05 eq. 4-HO-TEMPO, NaOCl
CO2, H2O, pH<7, <10 °C
44%
H
CO2H

In this oxidation the pH is adjusted below 7 by bubbling CO_2 into the sodium hypochlorite solution. The reaction fails when it is performed at a pH above 7, due to the sensitivity of propargyl alcohol to NaOCl under basic conditions. It is possible to increase the yield to 50–66% by performing a continuous oxidation.[27]

Selectivity

One of the most useful features of the oxidation of primary alcohols to carboxylic acids under Anelli's conditions is the great selectivity achieved for the oxidation of primary alcohols versus secondary ones.[8, 9b, 34] This selectivity is the result of the steric hindrance around the oxoammonium functionality in the oxoammonium salts derived from TEMPO-like radicals, resulting in a much easier attack by the relatively less hindered primary alcohols.

Research needed

The employment of amino oxyl radicals, yielding oxoammonium salts with a greater steric hindrance than those derived from TEMPO-like radicals, in the selective oxidation of primary alcohols must be investigated.

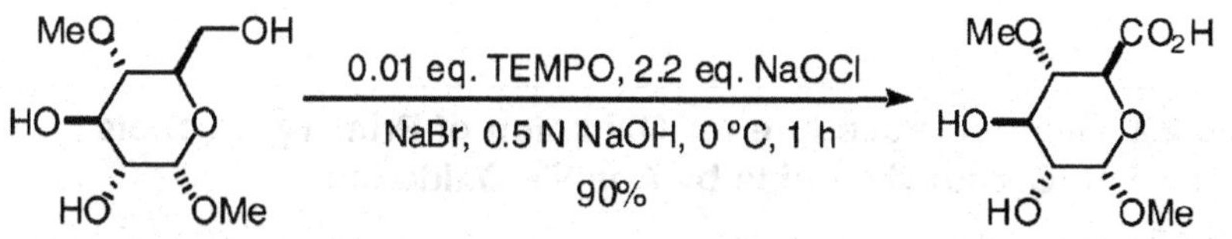

Treatment of the starting triol under Anelli's conditions in water at pH 10–11 allows the selective oxidation of the primary alcohol in 90%. The use of gaseous oxygen in the presence of platinum on carbon under Heyns' conditions provides a more modest 50% yield of the desired carboxylic acid.[9a]

Under basic conditions there is a compact and sterically demanding five-membered transition state for the oxidation of alcohols with oxoammonium salts, while under acidic conditions a less sterically demanding linear transition state operates. This results in both greater oxidizing speed and greater selectivity for oxidation of primary alcohols under Anelli's conditions at high pH. Furthermore, under less basic conditions hypohalous acids may compete with oxoammonium salts for the direct oxidation of alcohols[6b] resulting in even less selectivity. In fact, under acidic conditions the selectivity greatly decreases and secondary alcohols may even be oxidized more rapidly.[6a]

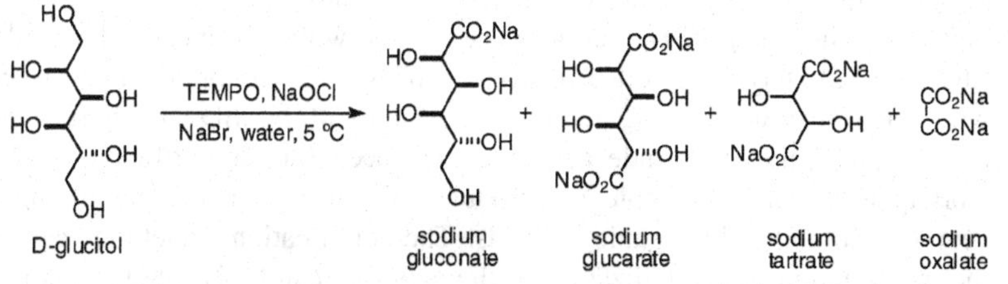

	sodium gluconate	sodium glucarate	sodium tartrate	sodium oxalate
pH= 10.5	24%	27%	18%	14%
pH= 11.7	1%	89%	5%	3%

It is possible to oxidize selectively the two primary alcohols in D-glucitol in 89% under Anelli's conditions at pH 11.7, with minor formation of side compounds resulting from incomplete oxidation or oxidative breakage at the secondary alcohols. Lowering of the pH to 10.5 produces a much less selective oxidation, resulting in only 27% of the desired diacid that is isolated as the disodium salt.[34b]

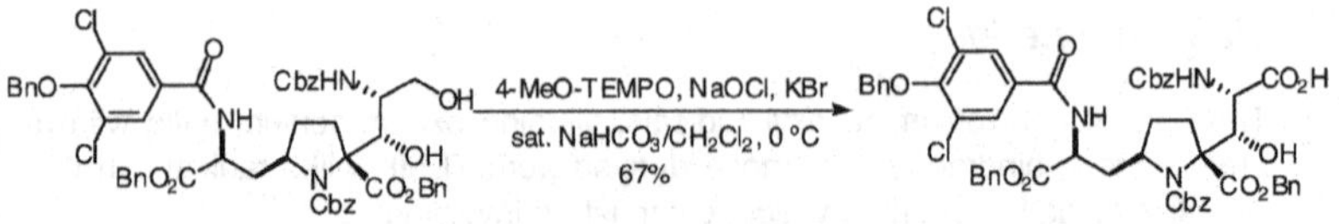

A primary alcohol is selectively oxidized in a complex molecule containing a secondary alcohol, using Anelli's conditions in a CH_2Cl_2–water biphasic system.[26]

6.2.1 General Procedure for Oxidation of Primary Alcohols to Carboxylic Acids by Anelli's Oxidation

A mixture of 1 equivalent of alcohol with ca. 0.002–0.1—typically 0.01—equivalent of TEMPO (MW 156.25),[a,b] ca. 1.75–8—typically 2—equivalents of NaOCl (MW 74.44),[c] ca. 0.1–0.7 equivalent of KBr (MW 119.01),[d] and ca. 0.05–0.12—typically 0.06—equivalent of a phase-transfer catalyst[e] in a biphasic buffered[f,g] system containing water and CH_2Cl_2,[h] is vigorously stirred at 0 °C[i] until most of the starting alcohol is consumed.[j] The reaction mixture is optionally quenched by addition of methanol or ethanol. When the reaction is carried out using a mixture of acetone and water, the removal of acetone *in vacuo* may facilitate the rest of the workup. The elimination of acetone sometimes causes the crystallization or precipitation of the sodium salt of the acid that can be isolated by filtration. Generally, the remainder of the workup of the reaction can be made according to three alternative protocols:

Workup A: The reaction mixture is optionally washed with an organic solvent like Et_2O or CH_2Cl_2. When the product—even as a sodium carboxylate—is very lipophilic, it may be convenient to adjust the pH above 12 by the addition of concentrated NaOH before the washing with an organic solvent. The pH is brought to ca. 2–6 by the addition of hydrochloric acid or 10% aqueous citric acid. WARNING: this may cause the evolution of poisonous chlorine gas. The mixture is extracted with an organic solvent like EtOAc or CH_2Cl_2. The collected organic phases can be optionally washed with water or brine. The organic phase is dried with Na_2SO_4 or $MgSO_4$ and concentrated, giving a crude acid that may need further purification. This workup may not be suitable for products like many sugars, possessing a very low solubility in organic solvents. The acidification sometimes causes the precipitation or crystallization of the acid that can be isolated by simple filtration. Alternatively, in the case of organic acids with a high solubility in water, the precipitation of the organic acid from an acidic aqueous phase may be induced sometimes by addition of an organic solvent miscible with water such as ethanol.

Workup B: The reaction mixture is fractioned between EtOAc and water. The organic phase is separated and can be optionally washed with water, 10% HCl, or brine. The organic phase is dried with Na_2SO_4 or $MgSO_4$ and concentrated, giving a crude acid that may need further purification. This workup is suitable for lipophilic organic acids possessing a very low solubility in water even when present as sodium salts.

Workup C: The reaction mixture is optionally neutralized by the addition of hydrochloric acid. The solvent is removed either by concentration at reduced pressure or lyophilization. The resulting residue containing the desired acid—either free or as a sodium salt contaminated with inorganic salts—is purified by chromatography.

[a] Other TEMPO derivatives such as 4-AcO-TEMPO (MW 214.28), 4-HO-TEMPO (MW 172.24), or 4-AcNH-TEMPO (MW 213.30) are equally effective.

[b] A quantity as high as 1–1.5 equivalents of TEMPO is sometimes employed in order to mitigate some side reactions induced by the stoichiometric oxidant, such as unwanted chlorinations.[22a,b, 28]

[c] Sodium hypochlorite is sold as a ca. 3–13% aqueous solution (chlorine bleach) containing some NaOH to provide a pH of 11–13, which helps to stabilize the reagent against disproportionation in NaCl and $NaClO_3$. Household bleach is perfectly effective.

[d] KBr is added to generate HOBr, which is a more effective secondary oxidant than HOCl. Failure to add KBr leads to a slower oxidation that nevertheless may prove useful. NaBr is equally effective.

[e] Quaternary ammonium salts such as n-Bu$_4$NCl (MW 277.9),[29] n-Bu$_4$NBr (MW 322.37),[30] n-Bu$_4$NHSO$_4$ (MW 339.53),[19] or Aliquat® 336 (tricaprylylmethylammonium chloride, MW 404.17)[31] are used as phase-transfer catalysts. Failure to add a phase-transfer catalyst results in a much slower reaction that may lead to isolation of aldehyde. No phase-transfer catalyst is needed when the oxidation is performed in a monophasic system.

[f] The addition of bleach causes the mixture to reach a very basic pH that may prove detrimental for many substrates. Thus, although the oxidation velocity increases under basic conditions, the pH is normally lowered to ca. 8.5–10 in order to attain a good balance of oxidation speed versus base-induced deleterious side reactions. The optimum pH is very substrate dependent. When the selective oxidation of a primary alcohol in the presence of a secondary one is desired, a pH as high as 11.7 may be advisable.[34b] On the other hand, in oxidations occurring in the presence of esters the pH must be lowered to 8–9 to avoid unwanted hydrolysis.

[g] The pH is normally lowered to 8.6–10 by the addition of an aqueous solution of $NaHCO_3$ possessing a concentration between 5% and saturation. A higher pH can be adjusted by the subsequent addition of aqueous NaOH. A final fine tuning of the pH can require the addition of some HCl. A phosphate buffer is sometimes used. The formation of carboxylic acid in the course of the oxidation leads to lowering of pH as the reaction proceeds, causing a decrease in the oxidation rate. This can be avoided by the sequential addition of some aqueous NaOH.

[h] A ca. 7:2 to 3:5 water:CH_2Cl_2 mixture is normally employed. Normally, ca. 15–60 mL of solvent mixture is used per mmol of alcohol. Sometimes, ca. 2–19 mL of brine per mmol of alcohol is also added. The reaction can also be performed in a monophasic system consisting of water, water:acetone (ca. 3:4 to 7:4), or water:MeCN (ca. 29:26 to 9:5). In a water:CH_2Cl_2 biphasic system the oxidation takes place in the organic phase, therefore, this system is very suitable for lipophilic substrates but fails in substrates—like most sugars—possessing a high solubility in water. In such cases the oxidation is best carried out in water or in a monophasic solvent mixture.

[i] The primary oxidants, which consist of oxoammonium salts, are very quickly decomposed by water at room temperature, leading to an actual decrease of oxidation rate with increasing temperature. Therefore, the reaction temperature must be kept close to $0\,^{\circ}C$ during the mixing of the reagents as soon as oxoammonium salts are generated. This may demand the slow addition of some reagents.

[j] It normally takes between 30 minutes and 20 hours.

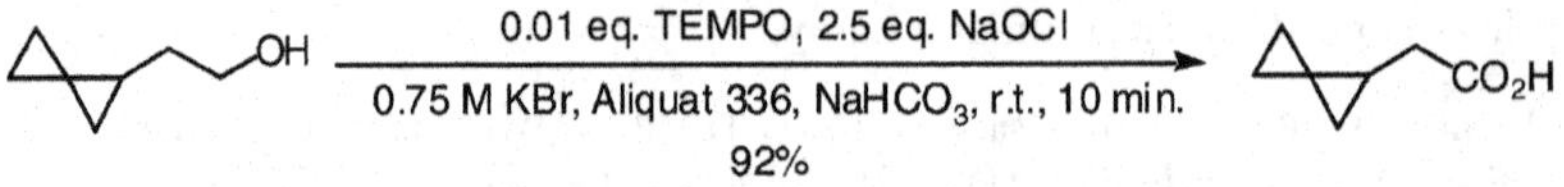

An attempted oxidation using Jones reagent resulted in extensive decomposition due to the acid-sensitivity of the spiro system. The employment of Anelli's oxidation under mildly basic conditions allowed the isolation of the desired acid in very good yield.[31a]

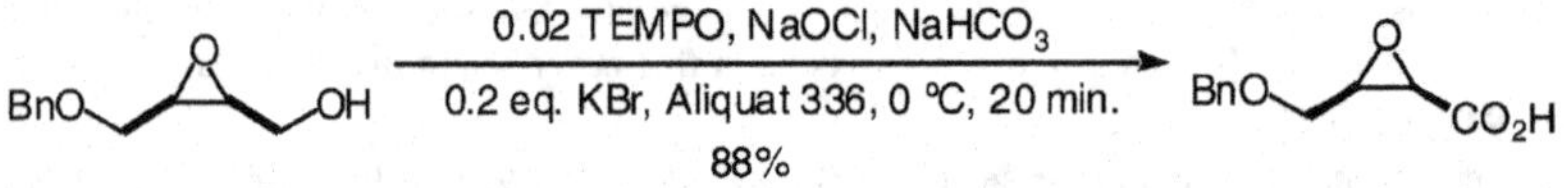

According to the authors: "Oxidation of the hydroxymethyl function to a carboxylic acid, without simultaneous oxidation of the benzyl to a benzoyl group, proved to be a challenging problem. Neither pyridinium dichromate in dimethylformamide nor Jones oxidation gave the desired product in acceptable yield. Oxidation with ruthenium tetraoxide predictably gave a mixture of the benzyl- and the benzoylepoxy acids. The problem was overcome by oxidation with hypochlorite in the presence of catalytic amounts of oxoammonium salt generated from 2,2,6,6-tetramethylpiperidine-1-oxyl (TEMPO, free radical). The product, (2R, 3R)-3-(benzyloxy)methyloxirane-2-carboxylic acid, was obtained in excellent yield (88%) and high purity."[31b]

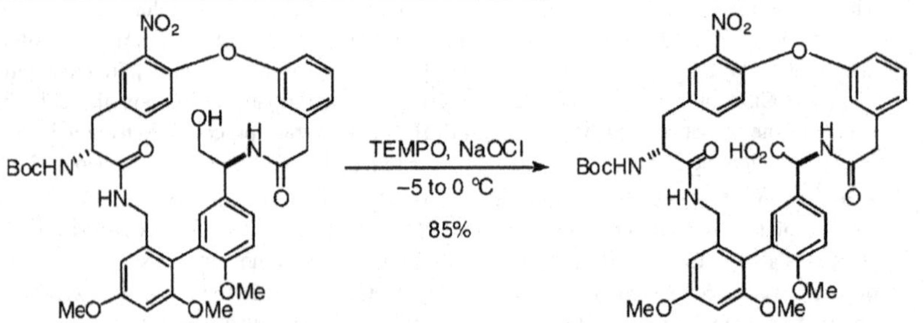

According to the authors: "Problems were encountered in the attempted oxidation of the primary alcohol to a carboxylic acid. After screening of various oxidation methods, the oxoammonium salt mediated oxidation using a couple TEMPO-NaOCl was selected for further optimization. Chlorination of the electron-rich aromatic ring was found to be the major side reaction. However, when the reaction temperature was carefully maintained between $-5\,^{\circ}C$ and $0\,^{\circ}C$, the desired acid was obtained in good yield."[35]

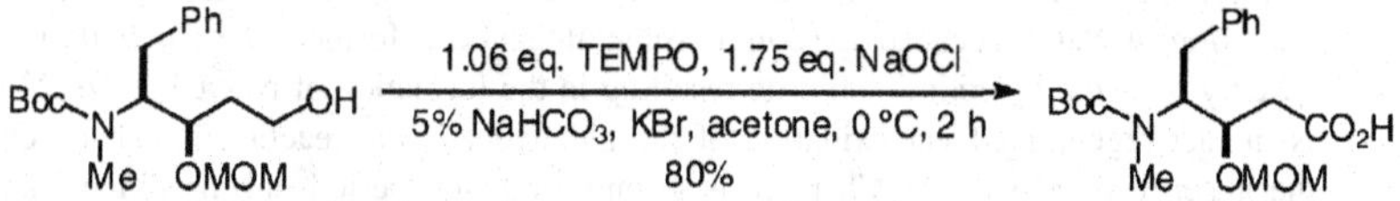

This oxidation under Anelli's conditions in which a quantity of TEMPO as high as 1.06 equivalents is utilized, probably to avoid side reactions, proved to be superior than the PDC/DMF system.[22b]

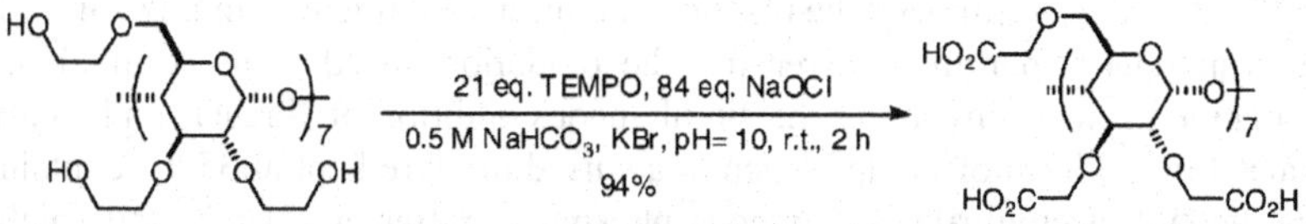

In this remarkable transformation, a total of 21 primary alcohols in one molecule are oxidized to carboxylic acids in a 94% yield, which means that each individual alcohol oxidation is performed in greater than 99.7% yield.[32]

6.3. Zhao's Modification of Anelli's Oxidation

In 1999, Zhao *et al.* published[4] a variation of Anelli's procedure for the transformation of primary alcohols into carboxylic acids, in which side reactions induced by the presence of sodium hypochlorite were lessened by using this reagent in catalytic rather than stoichiometric quantity. In this modified procedure sodium chlorite ($NaClO_2$) is employed as stoichiometric oxidant, which serves both to regenerate NaOCl and to operate as the primary oxidant[36] for the transformation of the intermediate aldehyde into carboxylic acid. The mechanism represented in the following scheme indicates the catalytic cycles in this oxidation:

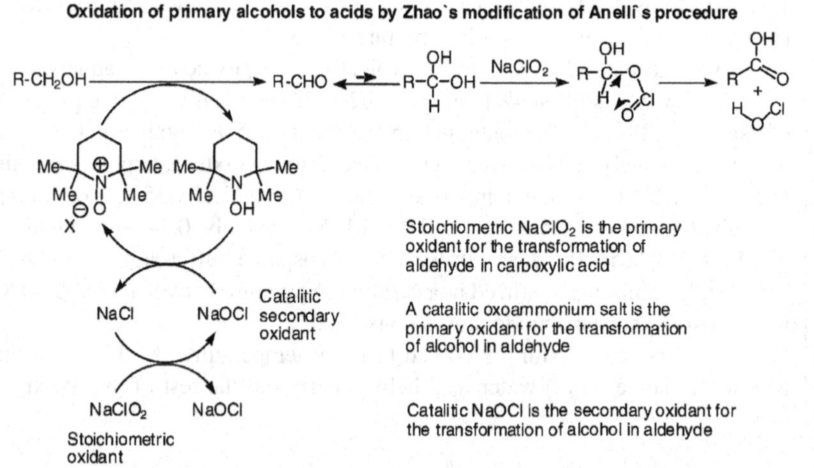

Oxidation of primary alcohols to acids by Zhao's modification of Anelli's procedure

Stoichiometric $NaClO_2$ is the primary oxidant for the transformation of aldehyde in carboxylic acid

A catalitic oxoammonium salt is the primary oxidant for the transformation of alcohol in aldehyde

Catalitic NaOCl is the secondary oxidant for the transformation of alcohol in aldehyde

Interestingly, there is no need for sodium chlorite to regenerate directly NaOCl by oxidation of NaCl, because as soon as some aldehyde is formed, the aldehyde is very quickly oxidized by sodium chlorite resulting in the formation of NaOCl. Thus, NaOCl is in fact regenerated via oxidation of the aldehyde. If the reaction is carried out in the absence of added NaOCl, there is a long induction period during which $NaClO_2$ generates the aldehyde in a very inefficient way. Once some aldehyde is formed, it is very quickly oxidized by $NaClO_2$, resulting in the formation of NaOCl that is very efficient in the generation of an N-oxoammonium salt, resulting in a very quick acceleration of the whole oxidation.

Zhao *et al.* optimized the reaction conditions seeking to minimize unwanted chlorinations rather than increasing the oxidation speed. This resulted in an oxidation protocol involving the simultaneous addition of $NaClO_2$ and catalytic NaOCl in the form of dilute bleach to a stirred mixture kept at 35 °C, containing the alcohol, acetonitrile, an aqueous phosphate buffer at pH 6.7, and catalytic TEMPO. WARNING: sodium chlorite and bleach must not be mixed before being added to the reaction, because the resulting mixture is unstable. Subsequent authors tended to follow very closely the original oxidation protocol of Zhao *et al.*

Zhao's modification of Anelli's oxidation is reported[4] to give generally better yields of carboxylic acids than the original Anelli's procedure. WARNING: Zhao's procedure involves the use of stoichiometric $NaClO_2$, which is a very powerful oxidant that can explode in the presence of organic matter. Therefore, Zhao's procedure must be employed, particularly on a big scale, only on substrates for which it proves to be clearly superior.

6.3.1 General Procedure for Oxidation of Primary Alcohols to Carboxylic Acids by Zhao's Modification of Anelli's Oxidation

WARNING: $NaClO_2$ and NaOCl must not be mixed before being added to the reaction because the resulting mixture is unstable.

Approximately, from 2 to 5—typically 2—equivalents of aqueous ca. 1.1–2 M $NaClO_2$ (MW 90.44)[a] and ca. 0.02–0.32—typically 0.02—equivalent of NaOCl (MW 74.44) contained in ca. 0.28–0.65%—typically 0.30%—bleach[b] are slowly[c] added over a vigorously stirred mixture, kept at a certain temperature between room temperature and 45 °C,[d] typically 35 °C, containing 1 equivalent of the alcohol, ca. 0.07–0.125—typically 0.1—equivalent of TEMPO (MW 156.25), acetonitrile,[e] and a phosphate buffer at pH 6.6–6.8.[f,g] The resulting mixture is stirred at a certain temperature between 35 and 50 °C until most of the starting alcohol is consumed.[h]

The reaction mixture is cooled to room temperature. At this point the addition of some (cold) water may help to carry out the rest of the workup.

The reaction mixture is basified to pH 8–9 by the addition of 0.05–2 M NaOH, and is quenched by mixing with a cold sodium sulfite (Na_2SO_3) aqueous solution, followed by stirring for 30 minutes. The mixture is washed with an organic solvent such as methyl t-butyl ether (MTBE), EtOAc, or Et_2O. The organic solutions must be checked for the presence of product, because very lipophilic organic acids can be washed away by organic solvents even from very basic aqueous solutions. The aqueous phase is acidified to a pH of 2–4 by the addition of 1–2 M HCl and extracted with an organic solvent such as EtOAc, Et_2O, MTBE, or CH_2Cl_2. If the organic phase from the washing of the aqueous basic solution contains product, all organic phases must be united for further workup. After optionally washing the organic phase containing the product with water and brine, it is dried (Na_2SO_4 or $MgSO_4$) and concentrated, giving a crude acid that may need further purification. The workup can be simplified by avoiding the quenching with sodium sulfite and the washing of the basified aqueous solutions with an organic solvent. This may lead to a more impure crude product.

[a] WARNING: $NaClO_2$ can explode in contact with organic matter.

[b] Household chlorine bleach is perfectly effective. The desired concentration of NaOCl is normally attained by diluting commercial bleach with some water.

[c] The $NaClO_2$ and NaOCl solutions can be added either simultaneously or sequentially beginning with $NaClO_2$. The addition is normally performed over a period between 15 minutes and 2 hours.

[d] Zhao *et al.* recommend a temperature of 35 °C. Other researchers utilize a slightly higher temperature of ca. 45–50 °C. It must be mentioned that the oxoammonium salts primary oxidants are unstable in hot water and increasing the temperature may lead to decreased reaction rate.

[e] A quantity of ca. 2.2–11.5—typically 5—mL of acetonitrile per mmol of alcohol is normally employed.

[f] An amount of ca. 1.7–5.5—typically 3.75—mL of phosphate buffer per mmol of alcohol is normally used.

[g] A phosphate buffer at pH 6.8 (35 °C) consists of an aqueous solution containing 0.025 M Na_2HPO_4 and 0.025 M KH_2PO_4.

[h] It normally takes between 4 and 24—typically 5—hours.

t-Bu ⏤ N(Boc) ⏤ OH   →   0.1 eq. TEMPO, 0.02 eq. NaOCl, 1,9 eq. $NaClO_2$ / MeCN/sodium phosphate buffer, 35 °C, 18 h / 84%   → t-Bu ⏤ N(Boc) ⏤ CO_2H

Although the oxidation could be carried out in 86% yield using $RuCl_3$ and $NaIO_4$, Zhao's modification of Anelli's procedure was preferred on a big scale.[37]

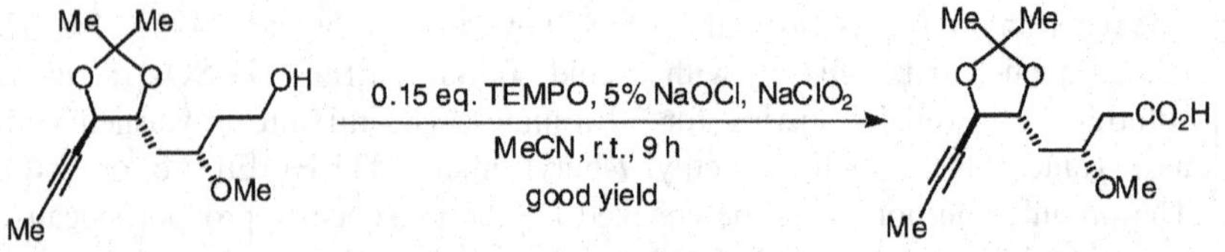

0.15 eq. TEMPO, 5% NaOCl, NaClO$_2$

MeCN, r.t., 9 h

good yield

A good yield of acid is obtained with catalytic TEMPO in the presence of stoichiometric NaClO$_2$ and catalytic NaOCl; other oxidations like PDC/DMF or Dess-Martin followed by NaClO$_2$ were found to be less effective.[38]

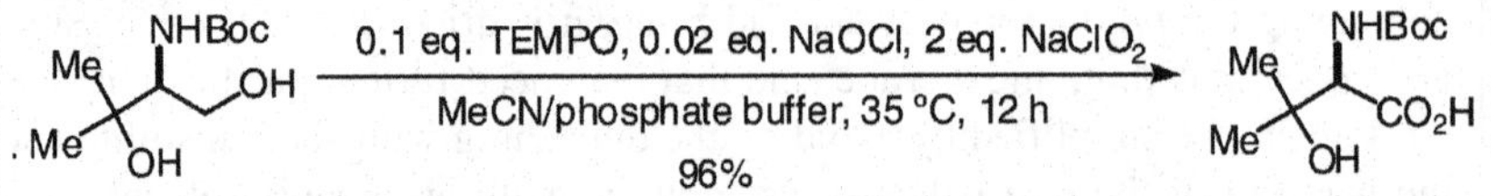

0.1 eq. TEMPO, 0.02 eq. NaOCl, 2 eq. NaClO$_2$

MeCN/phosphate buffer, 35 °C, 12 h

96%

While employing Zhao's modification of Anelli's oxidation leads to a 96% yield of carboxylic acid, a more expensive oxidation with oxygen in the presence of platinum results in a 76% yield.[39]

6.4. Oxidation of Epp and Widlanski

In 1997, Margarita, Piancatelli *et al.* reported[6c] the oxidation of alcohols to *aldehydes* and ketones using catalytic TEMPO in the presence of 1.1 equivalents of bis(acetoxy)iodobenzene (BAIB)—that is, PhI(OAc)$_2$—as stoichiometric secondary oxidant. According to the authors, under those conditions no noticeable overoxidation to carboxylic acid is detected, even when the reaction is carried out without employing an inert atmosphere and a dry solvent. In fact, Margarita and Piancatelli described one successful oxidation to aldehyde in which a MeCN:water (1:1) mixture is utilized as solvent. Two years later, Epp and Widlanski reported[5] that it is possible to obtain a carboxylic acid from a primary alcohol using the procedure of Margarita and Piancatelli, when the reaction is performed with at least 2 equivalents of BAIB in the presence of excess of water. They reported a successful protocol for the oxidation of primary alcohols to carboxylic acids involving 2.2 equivalents of BAIB and catalytic TEMPO in MeCN:water (1:1). This oxidation procedure for the obtention of carboxylic acids is rather unique, because it is carried out in the total absence of any inorganic salt and the by-products are the rather innocuous iodobenzene and acetic acid.

Mechanism

A mechanism consistent with the experimental facts is shown in the following scheme.

Oxidation of primary alcohols to acids by the procedure of Epp and Widlanski

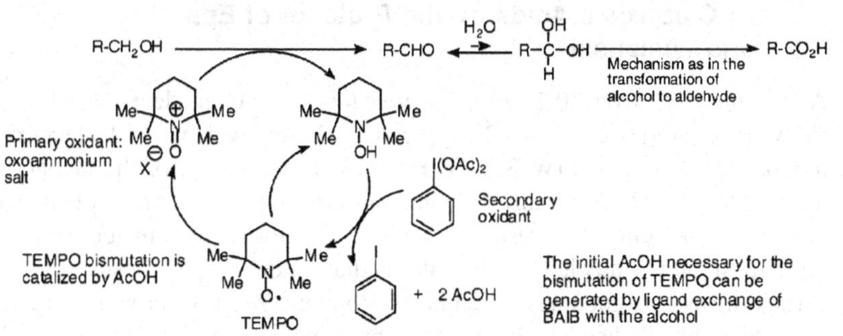

Apparently, BAIB does not oxidize directly TEMPO to the corresponding oxoammonium salt primary oxidant. Rather, TEMPO suffers an AcOH-catalyzed bismutation to an oxoammonium salt and hydroxylamine, the latter being oxidized to TEMPO by BAIB, resulting in the generation of AcOH. The initial AcOH necessary for the formation of the first molecules of oxoammonim salt can be generated by ligand exchange of PhI(OAc)$_2$ with the alcohol.

Selectivity

Although at the time of this writing the reports of selective oxidation of primary alcohols in the presence of secondary ones with TEMPO/BAIB are rather scarce,[40] it seems that this oxidation system shows a great potential for this selective transformation. The resulting hydroxyacids are very often isolated as lactones, and TEMPO/BAIB is a very good reagent for the conversion of 1,4- and 1,5-diols to lactones.[40b, 41]

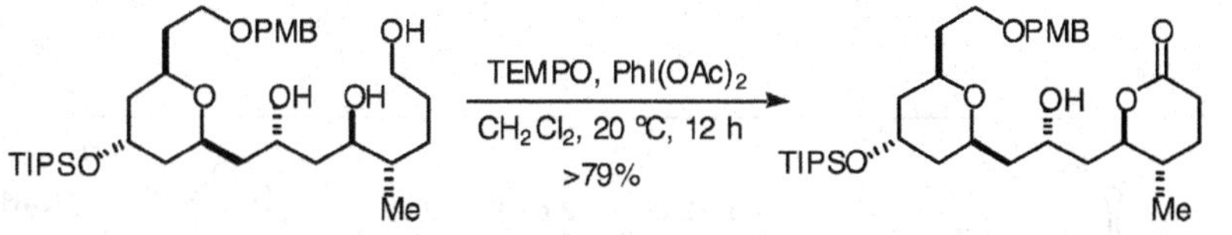

A primary alcohol is selectively oxidized in more than 79% in the presence of two secondary alcohols using BAIB with catalytic TEMPO. The resulting dihydroxyacid is isolated as a lactone.[41a]

6.4.1 General Procedure for Oxidation of Primary Alcohols to Carboxylic Acids by the Protocol of Epp and Widlanski

Approximately, from 0.2 to 0.3—typically 0.2—equivalent of TEMPO (MW 156.25) and ca. 2.2–4.2—typically 2.2—equivalents of bis(acetoxy) iodobenzene[a] (BAIB, MW 322.1) are added to a solution of the alcohol in a MeCN:water (1:1) mixture,[b,c] containing ca. 0.06–0.28 mmol of alcohol per mL of mixture. The resulting mixture is stirred at room temperature[d] until most of the alcohol or the intermediate aldehyde is consumed.[e] The carboxylic acid sometimes precipitates from the reaction mixture and can be very easily isolated in high purity by simple filtration. Otherwise, the reaction can be optionally quenched by addition of 5–10% aqueous $Na_2S_2O_3$ and the rest of the workup can proceed according to two alternative protocols:

Workup A:
The reaction mixture is concentrated, giving a crude acid containing iodobenzene[f] and other by-products that must be separated for example by chromatography.

Workup B:
The reaction mixture can be optionally acidified by the addition of hydrochloric acid. It is extracted with an aqueous solvent such as EtOAc or Et_2O. The organic phase can be optionally washed with water and brine. The resulting organic solution is dried (Na_2SO_4 or $MgSO_4$) and concentrated, giving a crude carboxylic acid that may need further purification.

[a] BAIB possesses the following ^{1}H-NMR (δ, $CDCl_3$, ppm): 8.13–7.45 (m, 5H, ArH), 2.01 (s, 6H, $MeCO_2$).[42]

[b] A CH_2Cl_2:water mixture can also be employed.[43]

[c] It is sometimes beneficial to add 1 or 2 equivalents of $NaHCO_3$ to the reaction mixture.[5]

[d] Better yields are sometimes obtained performing the reaction at 0 °C, a temperature at which the oxoammonium salt primary oxidant shows a greater stability in the presence of water.

[e] It normally takes between half an hour and 1 day.

[f] Iodobenzene possesses the following ^{1}H-NMR (δ, $CDCl_3$, ppm): 8.19 (s, 1H), 7.81 (s, 1H).[44]

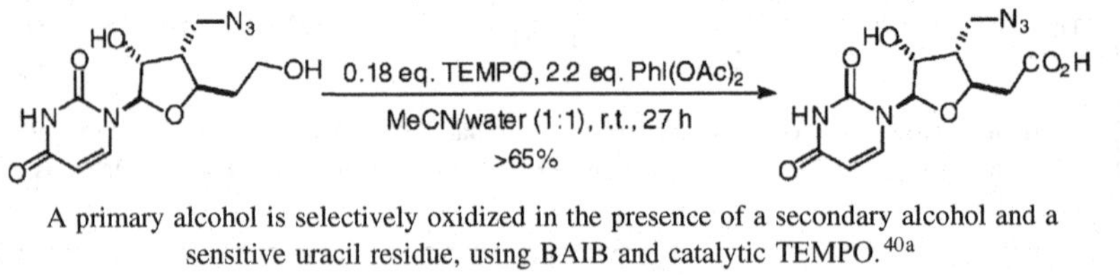

A primary alcohol is selectively oxidized in the presence of a secondary alcohol and a sensitive uracil residue, using BAIB and catalytic TEMPO.[40a]

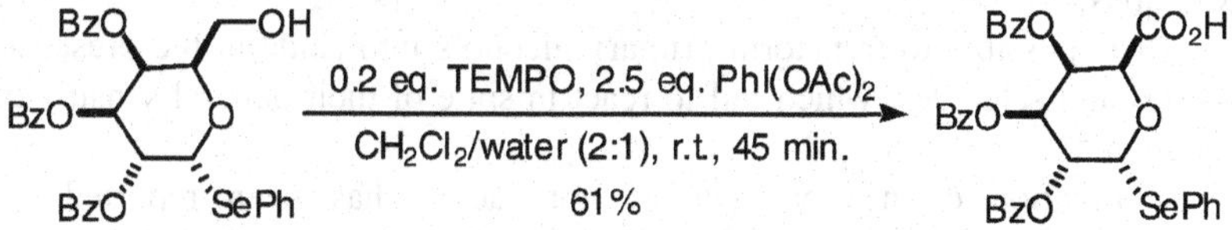

A primary alcohol is oxidized to carboxylic acid with BAIB and catalytic TEMPO in the presence of a very oxidation-sensitive selenoether.[40c]

6.5. Functional Group and Protecting Group Sensitivity to TEMPO-Mediated Oxidations

The transformation of primary alcohols into acids using Anelli's protocol involves more severe conditions than the analogous oxidation of alcohols to aldehydes and ketones.[22e]

TEMPO-mediated oxidations using Anelli's protocol are normally performed at a pH of ca. 8.6–10. This moderately basic pH is compatible with acid-sensitive functional groups and with most base-sensitive ones. Esters normally resist[40c, 45] the conditions of Anelli's oxidation, including the very base-sensitive acetate esters.[25] However, it is sometimes advisable to minimize the reaction time to prevent hydrolysis of acetates.[29a] Not surprisingly, esters are fully compatible with the almost neutral conditions of TEMPO-mediated oxidations by the method of Epp and Widlanski.[40c, 46]

Normally, Anellis's oxidation is not compatible with the presence of olefins, because usually they react very easily with NaOCl. Nevertheless, there is one report[22e] in which a primary alcohol is oxidized to acid in the presence of an alkene due to the use of stoichiometric—rather than catalytic—TEMPO that, therefore, is able to act as a chlorine scavenger.

Alkynes remain unchanged under TEMPO-mediated oxidations.[4, 47]

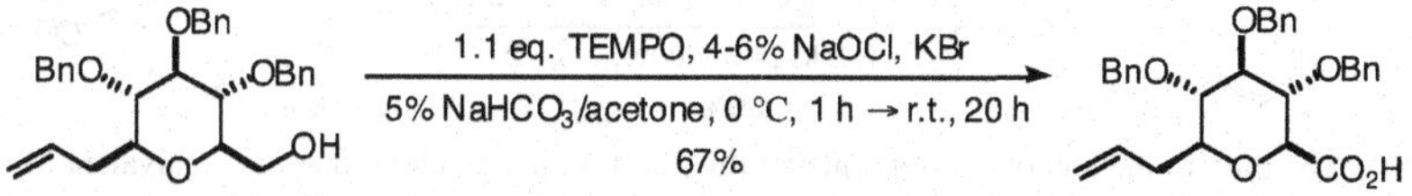

The oxidation of a primary alcohol to carboxylic acid under Anelli's conditions is performed using stoichiometric TEMPO that operates as a scavenger of chlorine that otherwise would react with the alkene.[22e]

Olefins conjugated with carbonyl groups are much less reactive against HOCl, and, therefore, remain unchanged under TEMPO-mediated oxidations.[34a, 48]

It is possible to transform primary alcohols into acids in the presence of sulfides and selenides, which fail to react in spite of their general sensitivity to oxidation.[40c]

Formation of neither lactones nor lactols has been reported when 1,4-[8, 9b, 34a, 49] and 1,5-diols[8, 14, 34b, 50] are oxidized under TEMPO-mediated conditions, unless the condition of Epp and Widlanski are used. In this case, good yields of lactones can be obtained.[40b, 41]

6.6. Side Reactions

A very common lateral reaction in oxidations under Anelli's protocol is the chlorination of aromatic rings and alkenes, due to the presence of HOCl. This reaction can be mitigated by:

- Addition of a stoichiometric quantity of TEMPO, which acts as scavenger of chlorine.[22a, 28c]
- Addition of a secondary oxidant different from NaOCl, such as t-BuOCl.[9b]
- Lowering the reaction temperature to -5 to $0\,^{\circ}$C.[35]
- Using Zhao's modification of Anelli's oxidation, in which NaClO$_2$—a much weaker chlorinator than HOCl—serves as stoichiometric secondary oxidant; HOCl being present in catalytic quantities and, therefore, much less able to produce undesired chlorinations.[4]

For obvious reasons, no chlorinations are possible in TEMPO-mediated oxidations performed under the protocol of Epp and Widlanski, in which PhI(OAc)$_2$ is used as secondary oxidant, rather than HOCl.

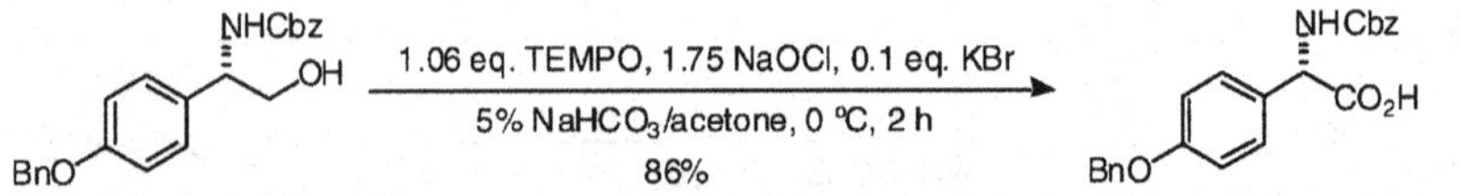

1.06 eq. TEMPO, 1.75 NaOCl, 0.1 eq. KBr

5% NaHCO$_3$/acetone, 0 $^{\circ}$C, 2 h

86%

According to the authors "In optimizing this reaction it was found that one equivalent of TEMPO was needed. When a catalytic amount of TEMPO was employed, a chlorinated aromatic derivative was isolated as the major product (presumably the TEMPO scavenges any chlorine which is liberated during the reaction provided that ca. 1 equiv is present)."[28c]

This oxidation is performed using t-BuOCl as secondary oxidant, rather than the more common HOCl, because the latter produces chlorination at the *para* position of the phenyl group. The similar benzyl and methyl glycosides can be oxidized uneventfully using HOCl as secondary oxidant.[9b]

Sometimes, carboxylic acids obtained by Anelli's oxidation suffer *in situ* decarboxylation. This can be avoided by proper adjustment of pH and maintaining low temperature.[29c]

According to the authors: "To prevent decarboxylation during the TEMPO oxidation, it was crucial to keep the pH between 8.5 and 9.5, and the temperature below 0 °C."[29c]

6.7. References

1 Golubev, V. A.; Rozantsev, E. G.; Neiman, M. B.; *Izv.Akad.Nauk SSSR Ser.Khim.* **1965**, *11*, 1927.

2 Cella, J. A.; Kelley, J. A.; Kenehan, E. F.; *J.Org.Chem.* **1975**, *40*, 1860.

3 Anelli, P. L.; Biffi, C.; Montanari, F.; Quici, S.; *J.Org.Chem.* **1987**, *52*, 2559.

4 Zhao, M.; Li, J.; Mano, E.; Song, Z.; Tschaen, D. M.; Grabowski, E. J. J.; Reider, P. J.; *J.Org.Chem.* **1999**, *64*, 2564.

5 Epp, J. B.; Widlanski, T. S.; *J.Org.Chem.* **1999**, *64*, 293.

6 a) de Nooy, A. E. J.; Besemer, A. C.; van Bekkum, H.; *Tetrahedron* **1995**, *51*, 8023. b) de Nooy, A. E. J.; Besemer, A. C.; van Bekkum, H.; *Carbohydr.Res.* **1995**, *269*, 89. c) De Mico, A.; Margarita, R.; Parlanti, L.; Vescovi, A.; Piancatelli, G.; *J.Org.Chem.* **1997**, *62*, 6974.

7 Dettwiler, J. E.; Lubell, W. D.; *J.Org.Chem.* **2003**, *68*, 177.

8 Ying, L.; Gervay-Hague, J.; *Carbohydr.Res.* **2003**, *338*, 835.

9 a) Li, K.; Helm, R. F.; *Carbohydr.Res.* **1995**, *273*, 249. b) Rye, C. S.; Withers, S. G.; *J.Am.Chem.Soc.* **2002**, *124*, 9756.

10 Semmelhack, M. F.; Schmid, C. R.; Cortés, D. A.; Chou, C. S.; *J.Am.Chem.Soc.* **1984**, *106*, 3374.

11 Inokuchi, T.; Matsumoto, S.; Nishiyama, T.; Torii, S.; *J.Org.Chem.* **1990**, *55*, 462.

12 Merbouh, N.; Bobbitt, J. M.; Brückner, C.; *J.Carbohydr.Res.* **2002**, *21*, 65.

13 De Luca, L.; Giacomelli, G.; Masala, S.; Porcheddu, A.; *J.Org.Chem.* **2003**, *68*, 4999.

14 Schnatbaum, K.; Schäfer, H. J.; *Synthesis* **1999**, 864.

15 Wu, A.; Yang, W.; Pan, X.; *Synth.Commun.* **1996**, *26*, 3565.

16 Heeres, A.; Van Doren, H. A.; Bleeker, I. P.; Gotlieb, K. F.; PCT Pat.Appl. WO 9636621 **1996**.

17 Siedlecka, R.; Skarżewski, J.; Młochowski, J.; *Tetrahedron Lett.* **1990**, *31*, 2177.
18 Ciriminna, R.; Blum, J.; Avnir, D.; Pagliaro, M.; *Chem.Commun.* **2000**, 1441.
19 Noula, C.; Loukas, V.; Kokotos, G.; *Synthesis* **2002**, 1735.
20 de Nooy, A. E. J.; Besemer, A. C.; van Bekkum, H.; *Recl.Trav.Chim. Pays-Bas* **1994**, *113*, 165.
21 See, for example: a) Brochette-Lemoine, S.; Joannard, D.; Descotes, G.; Bouchu, A.; Queneau, Y.; *J.Mol.Catal. A* **1999**, *150*, 31. b) Haller, M.; Boons, G.-J.; *J.Chem.Soc. Perkin Trans. I* **2001**, 814. c) Ibert, M.; Marsais, F.; Merbouh, N.; Brückner, C.; *Carbohydr.Res.* **2002**, *337*, 1059. d) Prabhu, A.; Venot, A.; Boons, G.-J.; *Org.Lett.* **2003**, *5*, 4975.
22 See, for example: a) Boger, D. L.; Borzilleri, R. M.; Nukui, S.; *J.Org.Chem.* **1996**, *61*, 3561. b) Pais, G. C. G.; Maier, M. E.; *J.Org.Chem.* **1999**, *64*, 4551. c) Hermann, C.; Giammasi, C.; Geyer, A.; Maier, M. E.; *Tetrahedron* **2001**, *57*, 8999. d) Boger, D. L.; Heon Kim, S.; Mori, Y.; Weng, J.-H.; Rogel, O.; Castle, S. L.; McAtee, J. J.; *J.Am.Chem.Soc.* **2001**, *123*, 1862. e) Xie, J.; *Eur.J.Org.Chem.* **2002**, 3411.
23 a) Bělohradský, M.; Císařová, I.; Holý, P.; Pastor, J.; Závada, J.; *Tetrahedron* **2002**, *58*, 8811. b) Bělohradský, M.; Buděšínský, M.; Císařová, I.; Dekoj, V.; Holý, P.; Závada, J.; *Tetrahedron* **2003**, *59*, 7751.
24 Rye, C. S.; Withers, S. G.; *J.Org.Chem.* **2002**, *67*, 4505.
25 Kraus, T.; Buděšínský, M.; Závada, J.; *Eur.J.Org.Chem.* **2000**, 3133.
26 Okue, M.; Kobayashi, H.; Shin-ya, K.; Furihata, K.; Hayakawa, Y.; Seto, H.; Watanabe, H.; Kitahara, T.; *Tetrahedron Lett.* **2002**, *43*, 857.
27 Stohrer, J.; Fritz-Langhals, E.; Bruninghaus, C.; Stauch, D.; U.S. Pat. 2003/0158439 **2003**.
28 a) Nicolaou, K. C.; Boddy, C. N. C.; Natarajan, S.; Yue, T.-Y.; Li, H.; Bräse, S.; Ramanjulu, J. M.; *J.Am.Chem.Soc.* **1997**, *119*, 3421. b) Nicolaou, K. C.; Ramanjulu, J. M.; Natarajan, S.; Bräse, S.; Li, H.; Boddy, C. N. C.; Rübsam, F.; *Chem.Commun.* **1997**, 1899. c) Reddy, K. L.; Sharpless, K. B.; *J.Am.Chem.Soc.* **1998**, *120*, 1207.
29 a) Davis, N. J.; Flitsch, S. L.; *Tetrahedron Lett.* **1993**, *34*, 1181. b) Brózda, D.; Koroniak, Ł.; Rozwadowska, M. D.; *Tetrahedron: Asymmetry* **2000**, *11*, 3017. c) Gruner, S. A. W.; Truffault, V.; Voll, G.; Locardi, E.; Stöckle, M.; Kessler, H.; *Chem.Eur.J.* **2002**, *8*, 4365. d) van Well, R. M.; Overkleeft, H. S.; van Boom, J. H.; Coop, A.; Wang, J. B.; Wang, H.; van der Marel, G. A.; Overhand, M.; *Eur.J.Org.Chem.* **2003**, 1704.
30 a) LePlae, P. R.; Umezawa, N.; Lee, H.-S.; Gellman, S. H.; *J.Org.Chem.* **2001**, *66*, 5629. b) Lefeber, D. J.; Aldaba Arévalo, E.; Kamerling, J. P.; Vliegenthart, J. F. G.; *Can.J.Chem.* **2002**, *80*, 76. c) Groth, T.; Meldal, M.; *J.Comb.Chem.* **2001**, *3*, 34.
31 a) Russo, J. M.; Price, W. A.; *J.Org.Chem.* **1993**, *58*, 3589. b) Wolf, E.; Spenser, I. D.; *J.Org.Chem.* **1995**, *60*, 6937.
32 Kraus, T.; Buděšínský, M.; Závada, J.; *J.Org.Chem.* **2001**, *66*, 4595.
33 Fraschini, C.; Vignon, M. R.; *Carbohydr.Res.* **2000**, *328*, 585.
34 See, for example: a) Bouktaib, M.; Atmani, A.; Rolando, C.; *Tetrahedron Lett.* **2002**, *43*, 6263. b) Thaburet, J.-F.; Merbouh, N.; Ibert, M.; Marsais, F.; Queguiner, G.; *Carbohydr.Res.* **2001**, *330*, 21.
35 Neuville, L.; Bois-Choussy, M.; Zhu, J.; *Tetrahedron Lett.* **2000**, *41*, 1747.
36 Lindgren, B. O.; Nilsson, T.; *Acta Chem.Scand.* **1973**, *27*, 888.
37 Halab, L.; Bélec, L.; Lubell, W. D.; *Tetrahedron* **2001**, *57*, 6439.
38 Smith III, A. B.; Minbiole, K. P.; Verhoest, P. R.; Schelhaas, M.; *J.Am.Chem.Soc.* **2001**, *123*, 10942.
39 Dettwiler, J. E.; Lubell, W. D.; *J.Org.Chem.* **2003**, *68*, 177.
40 a) Rozners, E.; Xu, Q.; *Org.Lett.* **2003**, *5*, 3999. b) Paterson, I.; Tudge, M.; *Tetrahedron* **2003**, *44*, 6833. c) van de Bos, L. J.; Codée, J. D. C.; van der Toorn, J. C.; Boltje, T. J.; Overkleeft, H. S.; van der Marel, G. A.; *Org.Lett.* **2004**, *6*, 2165.
41 a) Paterson, I.; Tudge, M.; *Angew.Chem.Int.Ed.* **2003**, *42*, 343. b) Hansen, T. M.; Florence, G. J.; Lugo-Mas, P.; Chen, J.; Abrams, J. N.; Forsyth, C. J.; *Tetrahedron Lett.* **2003**, *44*, 57.
42 Kazmierczak, P.; Skulski, L.; Kraszkiewicz, L.; *Molecules* **2001**, *6*, 881.

43 Raunkjær, M.; Bryld, T.; Wengel, J.; *Chem.Commun.* **2003**, *13*, 1604.
44 Rozen, S.; Zamir, D.; *J.Org.Chem.* **1990**, *55*, 3552.
45 Lefeber, D. J.; Kamerling, J. P.; Vliegenthart, J. F. G.; *Chem.Eur.J.* **2001**, *7*, 4411.
46 Raghavan, S.; Reddy, S. R.; *Tetrahedron Lett.* **2003**, *44*, 7459.
47 Koseki, Y.; Sato, H.; Watanabe, Y.; Nagasaka, T.; *Org.Lett.* **2002**, *4*, 885.
48 Desai, R. N.; Blackwell, L. F.; *Synlett* **2003**, 1981.
49 Söderman, P.; Widmalm, G.; *Eur.J.Org.Chem.* **2001**, 3453.
50 Kochkar, H.; Lassalle, L.; Morawietz, M.; Hölderich, W. F.; *J.Catal.* **2000**, *194*, 343.

7

Oxidation of Alcohols to Carboxylic Acids via Isolated Aldehydes

7.1. Introduction

Not surprisingly, the oxidation of primary alcohols to acids involves more rigorous experimental conditions than the oxidation to aldehydes, which possess a lower oxidation state than acids. Therefore, the milder oxidizing conditions needed for obtention of aldehydes offer better prospects for selective oxidations, particularly in sensitive and complex molecules possessing oxidation-sensitive moieties other than primary alcohols. Consequently, a two-step oxidation of primary alcohols to acids, via an intermediate aldehyde that is isolated, is a valuable synthetic alternative in some difficult substrates. The transformation of aldehydes into acids usually does not interfere with other sensitive functional groups because aldehydes are normally very easily oxidized under quite mild conditions. In fact, an inspection of the modern literature shows that, in a proportion as high as ca. 40%, the conversion of alcohols to acids is carried out using two separate synthetic operations, with isolation of the intermediate aldehydes. This highlights the fact that the direct oxidation of primary alcohols to acids is an immature technique in need of more selective reagents.

A listing of instances in which a two-step transformation of primary alcohols into acids is preferred over a direct one-step oxidation is shown below.

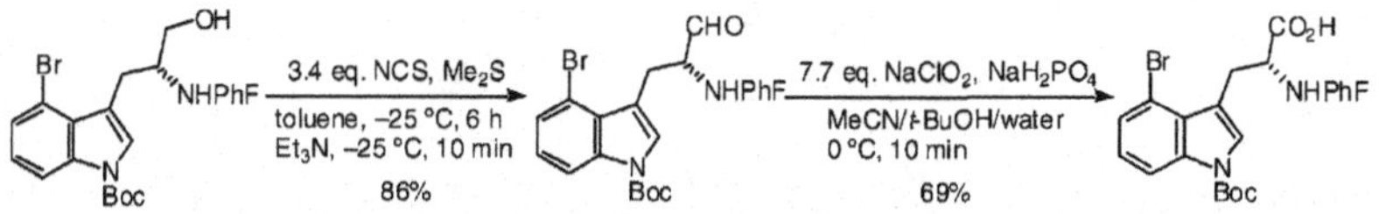

Initial attempts to oxidize directly the primary alcohol to carboxylic acid led to poor yields. A stepwise procedure secured the desired acid in a 59% overall yield.[1]

In this molecule possessing an inordinately complex structure, it was possible to achieve a remarkable 90% overall yield in the oxidation of a primary alcohol to acid, due to the use of a two-step oxidation via aldehyde. According to the authors: "Initially, direct conversion to the carboxylic acid was accomplished by using the Sharpless oxidation protocol employing ruthenium (IV) oxide as catalyst. However, the yield varied according to the scale used, and removal of the ruthenium salts was difficult. Therefore, a stepwise oxidation was chosen. Conversion of the alcohol to the aldehyde was accomplished by a Swern oxidation with trifluoroacetic anhydride as the DMSO activator. The unstable aldehyde was immediately oxidized to the carboxylic acid, by using a procedure developed by Masamune for oxygen-rich molecules containing acid-sensitive groups. Treatment of the aldehyde with potassium permanganate in t-butyl alcohol with use of 5% sodium hydrogen phosphate, gave, after 30 min, the carboxylic acid in 90% yield".[2]

The direct oxidation of the alcohol to acid was thwarted by the tendency of the alcohol to suffer an oxidative carbon–carbon bond scission at the hydroxyethyl side chain. For example, treatment with PDC resulted partially in removal of one carbon atom with generation of the corresponding arylcarboxylic acid. The desired acid could be secured by the combination of two very mild oxidations, consisting of the generation of the aldehyde by the Corey–Kim method, followed by treatment with Ag (I).[3]

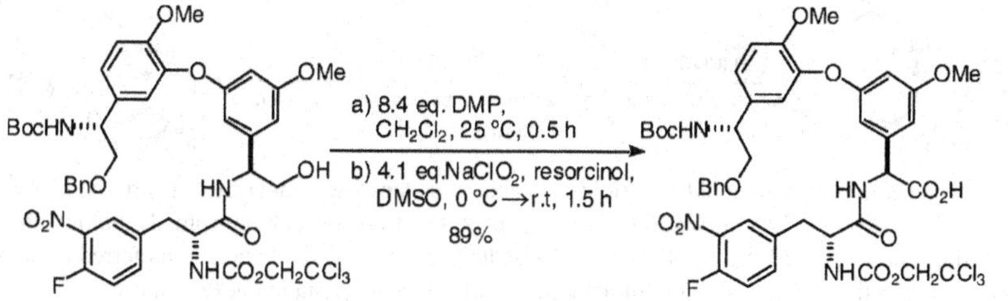

During the oxidation of the primary alcohol in this complex substrate, it was very difficult to avoid a very facile epimerization at the α position that occurred at the intermediate aldehyde stage. Eventually, it was possible to achieve the desired oxidation to acid, with little or no epimerization, by performing an oxidation to aldehyde with Dess–Martin periodinane, followed by treatment with $NaClO_2$ in the presence of resorcinol. The workup of the Dess–Martin oxidation had to be performed at $0\,°C$ to avoid epimerization. It was possible to obtain the acid by direct oxidation of the alcohol with TEMPO-NaOCl/KBr, also resulting in little or no epimerization; however, this method was not preferred, because it delivered lower and erratic conversions (0–43%).[4]

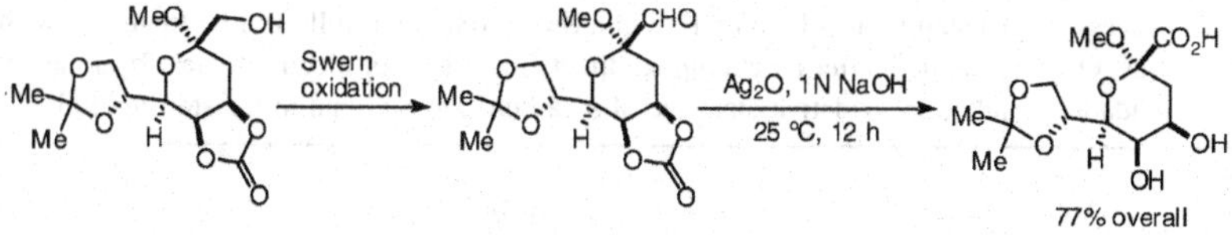

According to the authors: "Attempts to effect the direct oxidation of the C(1) hydroxyl to a carboxyl group failed, but a convenient stepwise procedure was devised that entailed Swern oxidation followed by oxidation of the intermediate aldehyde under conditions that proceed with concomitant hydrolysis of the carbonate moiety (Ag_2O, 1N NaOH, $25\,°C$, 12 h) to furnish the acid in 77% yield."[5]

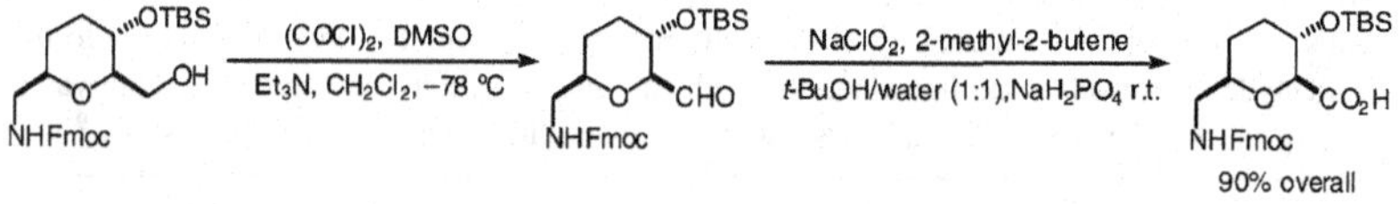

While a direct oxidation to acid using catalytic RuO_4 furnished the acid in 70% yield, a two-step oxidation via aldehyde, using Swern oxidation followed by $NaClO_2$, provided the acid in 90% overall yield.[6]

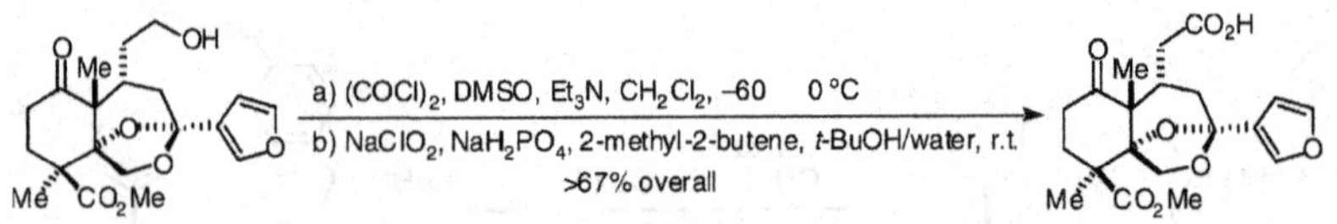

a) (COCl)$_2$, DMSO, Et$_3$N, CH$_2$Cl$_2$, –60   0 °C

b) NaClO$_2$, NaH$_2$PO$_4$, 2-methyl-2-butene, t-BuOH/water, r.t.

\>67% overall

The starting alcohol could be oxidized to acid in a stepwise manner by a Swern oxidation followed by NaClO$_2$. It was possible to make a direct transformation from alcohol to acid using Zhao's procedure (TEMPO-NaOCl-NaClO$_2$), but this method was not preferred because it delivered a lower yield of acid due to concomitant oxidation at the furan ring.[7]

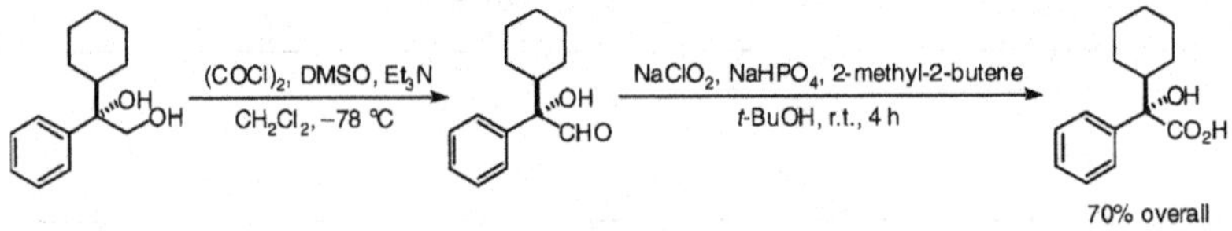

(COCl)$_2$, DMSO, Et$_3$N

CH$_2$Cl$_2$, –78 °C

NaClO$_2$, NaHPO$_4$, 2-methyl-2-butene

t-BuOH, r.t., 4 h

70% overall

A successful two-step protocol, consisting in a Swern oxidation followed by treatment with NaClO$_2$, was applied, after an attempted direct conversion of alcohol to acid by Jones oxidation resulted in oxidative cleavage of a carbon–carbon bond in the starting diol.[8]

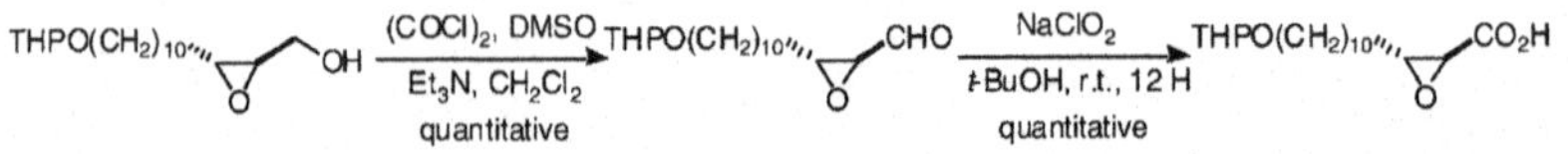

(COCl)$_2$, DMSO

Et$_3$N, CH$_2$Cl$_2$

quantitative

NaClO$_2$

t-BuOH, r.t., 12 H

quantitative

According to the authors: "The one-step oxidation procedure with RuO$_4$ was not possible in this case due to the vulnerability of the THP protecting group. The Swern oxidation to the aldehyde, followed by further oxidation with sodium chlorite to the carboxylic acid was a perfect alternative." The authors claimed a quantitative conversion in both steps.[9]

A direct oxidation to carboxylic acid using PDC in DMF was unsuccessful. Therefore, the authors opted [10] for a three-step procedure beginning with oxidation to aldehyde with PDC in CH_2Cl_2. This is followed by an interesting two-step oxidation of the aldehyde to acid via a methyl ester—obtained with PDC in DMF in the presence of methanol after O'Connor and Just [11]— that is hydrolyzed to acid under basic conditions.

7.2. References

1 Hurt, C. R.; Lin, R.; Rapoport, H.; *J.Org.Chem.* **1999,** *64,* 225.

2 Li, W.-R.; Ewing, W. R.; Harris, B. D.; Joullié, M. M.; *J.Am.Chem.Soc.* **1990,** *112,* 7659.

3 Sahali, Y.; Skipper, P. L.; Tannenbaum, S. R.; *J.Org.Chem.* **1990,** *55,* 2918.

4 Boger, D. L.; Heon Kim, S.; Mori, Y.; Weng, J.-H.; Rogel, O.; Castle, S. L.; McAtee, J. J.; *J.Am.Chem.Soc.* **2001,** *123,* 1862.

5 Martin, S. F.; Zinke, P. W.; *J.Am.Chem.Soc.* **1989,** *111,* 2311.

6 Aguilera, B.; Siegal, G.; Overkleeft, H. S.; Meeuwenoord, N. J.; Rutjes, F. P. J. T.; van Hest, J. C. M.; Schoemaker, H. E.; van der Marel, G. A.; van Boom, J. H.; Overhand, M.; *Eur.J.Org.Chem.* **2001,** 1541.

7 Boeckman Jr., R. K.; Rico Ferreira, M. R.; Mitchell, L. H.; Shao, P.; *J.Am.Chem.Soc.* **2002,** *124,* 190.

8 Gupta, P.; Fernandes, R. A.; Kumar, P.; *Tetrahedron Lett.* **2003,** *44,* 4231.

9 Thijs, L.; Zwanenburg, B.; *Tetrahedron* **2004,** *60,* 5237.

10 Koskinen, A.M.P.; Hassila, H.; Myllymäki, V.; Rissanen, K.; *Tetrahedron Lett.* **1995,** *36,* 5619.

11 O'Connor, B.; Just, G.; *Tetrahedron Lett.* **1987,** *28,* 3235.

Index